THE SAFETY AND HEALTH HANDBOOK

DAVID L. GOETSCH

Prentice Hall

Upper Saddle River, New Jersey
Columbus, Ohio

Library of Congress Cataloging-in-Publication Data
Goetsch, David L.
        The safety and health handbook / David L. Goetsch.
                p.   cm.
        Includes bibliographical references and index.
        ISBN 0-13-674243-2
        1. Industrial hygiene Handbooks, manuals, etc. 2. Industrial safety Handbooks, manuals,
etc.   I. Title
        RC967.G66   2000
        616.9′803—dc21
                                                                99-32911
                                                                CIP

Editor: Stephen Helba
Assistant Editor: Michelle Churma
Production Editor: Louise N. Sette
Design Coordinator: Karrie Converse-Jones
Cover Designer: Mark Shumaker
Cover art: © Steven Schildbach
Production Manager: Matthew Ottenweller
Electronic Text Management: Marilyn Wilson Phelps, Karen L. Bretz, Melanie King
Marketing Manager: Chris Bracken

This book was set in Clearface and Swiss by Prentice Hall and was printed and bound by The Banta Company. The cover was printed by Phoenix Color Corp.

Printed in the United States of America

10 9 8 7 6 5 4 3 2 1

ISBN: 0-13-674243-2

Prentice-Hall International (UK) Limited, *London*
Prentice-Hall of Australia Pty. Limited, *Sydney*
Prentice-Hall of Canada, Inc., *Toronto*
Prentice-Hall Hispanoamericana, S. A., *Mexico*
Prentice-Hall of India Private Limited, *New Delhi*
Prentice-Hall of Japan, Inc., *Tokyo*
Prentice-Hall (Singapore) Pte. Ltd., *Singapore*
Editora Prentice-Hall do Brasil, Ltda., *Rio de Janeiro*

Preface

BACKGROUND

Occupational safety and health are issues that have a direct bearing on an organization's performance and, correspondingly, its competitiveness. Consequently, these issues should be of interest to managers at all levels in organizations.

The field of occupational safety has undergone significant changes over the past two decades. There are many reasons for this. Some of the more prominent include the following: technological changes that have introduced new hazards in the workplace; proliferation of safety legislation and corresponding regulations; increased pressure from regulatory agencies; realization by executives that a safe workplace is typically a more productive workplace; skyrocketing health care and workers' compensation costs; increased pressure from environmental groups and the public; a growing interest in corporate responsibility; professionalization of safety occupations; increased pressure from labor organizations and employees in general; and rapidly mounting costs associated with product safety and other types of litigation. These factors have created a need for an up-to-date book that contains the fundamental information managers, technologists, and engineers need to know about occupational safety.

WHY WAS THIS BOOK WRITTEN AND FOR WHOM?

This book was written in response to the need for an up-to-date handbook that focuses on the fundamental information managers, engineers, and technologists need to know about occupational safety. It is intended for use in universities, colleges, community colleges, and corporate training settings that offer programs, courses, workshops, and/or seminars in occupational safety, the basics of safety, and/or safety management. The direct, straightforward presentation of material in this book focuses on making the theories and principles of occupational safety practical and useful in an actual work setting. Up-to-date research has been integrated throughout in a down-to-earth manner. Checklists and other timesaving devices are used frequently throughout.

ORGANIZATION OF THE BOOK

The text contains nineteen chapters, each focusing on a major area of concern for the modern workplace. The chapters are presented in an order that is compatible with the typical organization of a college-level safety course. A standard chapter format is used throughout the book. Each chapter begins with a list of major topics and ends with simulated scenarios that require the reader to apply what is being learned.

SPECIAL FEATURES OF THE BOOK

Today's managers, engineers, and technologists are busy people. The global workplace is competitive, constantly changing, and hectic. Consequently, this book was organized and written in such a way as to give the reader the maximum amount of information in the least amount of time. Features making information accessible and convenient include easy-to-use illustrations and checklists throughout each chapter. Bulleted items and checklists typically represent the most important information in the chapter. The illustrations are designed to plant a mental image of important information in the mind of the reader and to make the information visual so that it is more easily remembered.

A final feature is the Application Scenario. Instead of end-of-chapter questions that test the reader's retention of facts, real-world scenarios are provided that require the reader to apply the concepts presented in that chapter. They can be completed individually and/or in groups. Many of the scenarios are based on actual events that occurred in actual organizations. Solving the problems contained in these scenarios will give readers practice in dealing with the types of situations they are likely to confront in a live setting. A companion website, with interactive quizzes, links to safety and health websites, and lecture notes is available at www.prenhall.com/goetsch.

HOW THIS BOOK DIFFERS FROM OTHERS

Most textbooks on occupational safety and health are developed for students who are majoring in a safety and health-related discipline with the goal of becoming a full-time safety professional. The author's book, *Occupational Safety and Health for Technologists, Engineers, and Managers*, Third Edition, is such a text.

This handbook, however, is for individuals in other fields or in related fields who have an interest in or partial responsibility for safety. Individuals who need *handbook-level* knowledge as opposed to *practitioner's depth* knowledge will find this book helpful. Every manager, engineer, and technologist in any company needs to be sufficiently knowledgeable about safety and health to be an effective player in keeping the workplace hazard-free.

Students majoring in any management, engineering, or technology discipline will need to be knowledgeable about safety and health and how they affect productivity, quality, and competitiveness. This book provides that level of knowledge.

ABOUT THE AUTHOR

David L. Goetsch is Provost of the joint campus of the University of West Florida and Okaloosa-Walton Community College in Fort Walton Beach, Florida, and Professor of Safety and Quality Management. In addition, Dr. Goetsch is President of The Management Institute, a partnership of the University of West Florida, Okaloosa-Walton Community College, and the Okaloosa Economic Development Council. He is also the founder of the college's Institute for Professional Development, and the author of Prentice Hall's leading textbook on workplace safety: *Occupational Safety and Health for Technologists, Engineers, and Managers,* Third Edition.

Contents

CHAPTER ELEVEN
Radiation Hazards   146

CHAPTER TWELVE
Noise and Vibration Hazards   156

CHAPTER THIRTEEN
Automation and Technology Hazards   167

CHAPTER FOURTEEN
Bloodborne Pathogens   178

Accident Costs

There is a long history of debate in this country concerning the effect of accidents on industry (the workers and the companies) and the cost of preventing accidents. Historically, the prevailing view was that accident prevention programs were too costly. The more contemporary view is that accidents are too costly and that accident prevention makes sense economically. As a result, accident prevention, which had been advocated on a moral basis, is now justified in economic terms.

Accidents are the fourth leading cause of death in this country after heart disease, cancer, and strokes. This ranking is based on all types of accidents including motor vehicle accidents, drownings, fires, falls, natural disasters, and work-related accidents.

Although deaths from natural disasters tend to be more newsworthy than workplace deaths, their actual impact is substantially less. For example, the most publicized natural disasters in the United States typically cause fewer than 100 deaths annually. Workplace accidents account for approximately 10,400 accidental workplace deaths every year.[1] The following quote from the National Safety Council puts workplace accidents and deaths in the proper perspective, notwithstanding their apparent lack of newsworthiness.

> While you make a 10-minute speech—2 persons will be killed and about 170 will suffer a disabling injury. Costs will amount to $2,800,000. On the average, there are 11 accidental deaths and about 1,030 disabling injuries every hour during the year.[2]

This chapter provides the information needed to have a full understanding of workplace accidents and their effect on industry in the United States. Such an understanding will help professionals play a more effective role in keeping both management and labor focused appropriately on safety and health in the workplace.

COSTS OF ACCIDENTS

A news brief appearing in the *Occupational Health & Safety Letter* reported that an Illinois contracting firm was fined $750,000 on charges that it willfully ignored federal safety rules and caused the death of three workers in a sewer tunnel explosion in Milwaukee.[3]

To gain a proper perspective on the economics of workplace accidents, we must view them in the overall context of all accidents. The overall cost of accidents in the United States is approximately $150 billion. These costs include such factors as lost wages, medical expenses, insurance administration, fire-related losses, motor vehicle property damage, and indirect costs.

Figure 1–1 breaks down this overall amount by categories of accidents. Figure 1–2 breaks them down by cost categories. Notice in Figure 1–1 that workplace accidents rank second behind motor vehicle accidents in cost. Figure 1–2 shows that the highest cost category is wages lost by workers who are either injured or killed. The category of indirect losses from work accidents consists of costs associated with responding to accidents (i.e., giving first aid, filling out accident reports, handling production slowdowns).

Clearly accidents on and off the job cost U.S. industry dearly. Every dollar that is spent responding to accidents is a dollar that could have been reinvested in modernization, research and development, facility upgrades, and other competitiveness-enhancing activities.

ACCIDENTAL DEATHS IN THE UNITED STATES

Accidental deaths in the United States result from a variety of causes, including motor vehicle accidents, falls, poisoning, drowning, fire-related injuries, suffocation (ingested object), firearms, medical complications, air transport accidents, machinery, mechanical suffocation, and the impact of falling objects. The National Safety Council periodically computes death totals and death rates in each of these categories. The statistics for a typical year are as follows:

- *Motor vehicle accidents.* Motor vehicle accidents are the leading cause of accidental deaths in the United States each year. They include deaths resulting from accidents involving mechanically or electrically powered vehicles (excluding rail vehicles) that occur on or off the road. In a typical year, there are approximately 47,000 such deaths in the United States.

- *Falls.* This category includes all deaths from falls except those associated with transport vehicles. For example, a person who is killed as the result of falling while boarding a bus or train would not be included in this category. In a typical year, there are approximately 13,000 deaths in the United States from falls.

- *Poisoning.* The poisoning category is divided into two subcategories: (1) poisoning by solids and liquids, and (2) poisoning by gases and vapors. The first category includes deaths that result from the ingestion of drugs, medicine, widely recognized solid and liquid poisons, mushrooms, and shellfish. It does not include poisoning from spoiled food or salmonella. The second category includes death caused by incomplete combustion (for example, gas vapors from an oven or unlit pilot light) or from carbon monoxide (for example, exhaust fumes from an automobile). In a typical year, there are approximately 6,000 deaths in the first category and 1,000 in the second.

Figure 1–1
Accident costs by accident type (in billions, in a typical year).

Motor vehicle accidents	$722.0
Workplace accidents	48.5
Home accidents	18.2
Public accidents	12.5

Figure 1–2
Accident costs by categories (in billions, in a typical year).

Wages lost	$37.7
Medical expenses	23.7
Insurance administration	28.4
Property damage (motor vehicle)	26.8
Fire losses	9.4
Indirect losses for work accidents	22.5

- *Drowning.* This category includes work-related and nonwork-related drownings but excludes those associated with floods or other natural disasters. In a typical year, there are approximately 5,000 deaths from drowning in the United States.
- *Fire-related injuries.* This category includes deaths from burns, asphyxiation, falls, and those that result from falling objects in a fire. In a typical year, there are over 4,000 fire-related deaths in the United States.
- *Suffocation (ingested object).* This category includes deaths from the ingestion of an object that blocks the air passages. In many such deaths, the ingested object is food. In a typical year, there are approximately 4,000 suffocation deaths in the United States.
- *Firearms.* This category includes deaths that result when recreational activities involving firearms or household accidents involving firearms result in death. For example, a person killed in the home while cleaning a firearm would be included in this category. However, a person killed in combat would not be. In a typical year, there are approximately 2,000 deaths in this category.
- *Others.* This category includes deaths resulting from medical complications arising out of mistakes made by health care professionals, air transport injuries, interaction with machinery, mechanical suffocation, and the impact of falling objects. In a typical year, there are over 14,000 deaths in these subcategories.[4]

ACCIDENTS VS. OTHER CAUSES OF DEATH

Although there are more deaths each year from heart disease, cancer, and strokes than from accidents, these causes tend to be concentrated among people at or near retirement age. Among people 37 years of age or younger—prime working years—accidents are the number one cause of death. Figure 1–3 summarizes the causes of death for persons from 25 to 44 years of age. Notice that the leading cause is accidents.

Figure 1–3 shows that accidents represent a serious detriment to productivity, quality, and competitiveness in today's workplace. Yet accidents are the one cause of death and injury that companies can most easily control. Although it is true that companies

Figure 1–3
Causes of accidents (ages 25 to 44 years in a typical year).

Accidents	27,484
Motor vehicle	16,405
Poison (solid, liquid)	2,649
Drowning	1,516
Falls	1,138
Fire-related	899
Cancer	20,305
Heart disease	15,874

might have some success in decreasing the incidence of heart disease and stroke among their employees through such activities as corporate wellness programs, their impact in this regard will be limited. However, employers can have a significant impact on preventing accidents.

WORK ACCIDENT COSTS AND RATES

Workplace accidents cost employers millions every year. Consider the following example. Arco Chemical Company was ordered to pay $3.48 million in fines as a result of failing to protect workers from an explosion at its petrochemical plant in Channelview, Texas. The steel-making division of USX paid a $3.25 million fine to settle numerous health and safety violation citations. BASF Corporation agreed to pay a fine of $1.06 million to settle OSHA citations associated with an explosion at a Cincinnati chemical plant that caused two deaths and seventeen injuries.

These examples show the costs of fines only. In addition to fines, these employers incurred costs for safety corrections, medical treatment, survivor benefits, death/burial costs, and a variety of indirect costs. Clearly, work accidents are expensive. However, the news is not all bad. The trend in the rate of accidents is downward.

Work accident rates in this century are evidence of the success of the safety movement in the United States. As the amount of attention given to workplace safety and health has increased, the accident rate has decreased.[5]

As was shown in Figure 1–1, the cost of these 10,000 work deaths and work injuries in a typical year is $48.5 billion. This translates into a cost of $420 per worker in the United States, computed as the value-added required per worker to offset the cost of work injuries. It translates further into $610,000 per death and $18,000 per disabling injury.[6]

Although statistics are not available to document the supposition, many safety and health professionals believe that the major cost of accidents and injuries on the job results from damage to morale. Employee morale is less tangible than documentable factors such as lost time and medical costs. However, it is widely accepted among management professionals that few factors affect productivity more than employee morale. Employees with low morale do not produce up to their maximum potential. This is why so much time and money are spent every year to help supervisors and managers learn different ways to help improve employee morale.

Since few things are so detrimental to employee morale as seeing a fellow worker injured, accidents can have a devastating effect. Whenever an employee is injured, his or her colleagues silently think, "That could have been me," in addition to worrying about the employee. Morale is damaged even more if the injured employee is well liked and other employees know his or her family.

TIME LOST BECAUSE OF WORK INJURIES[7]

An important consideration when assessing the effect of accidents on industry is the amount of lost time due to work injuries. According to the National Safety Council, approximately 35,000,000 hours are lost in a typical year as a result of accidents. This is actual time lost from disabling injuries and does not include additional time lost for medical checkups after the injured employee returns to work. Accidents that occurred in previous years often continue to cause lost time in the current year.

DEATHS IN WORK ACCIDENTS

Deaths on the job have decreased markedly over the years. However, they still occur. For example, in a typical year, there are 10,400 work deaths in the United States. The causes of death in the workplace vary. They include those related to motor vehicles, falls, electric current, drowning, fires, air transport, poison, water transport, machinery, falling objects, rail transports, workplace violence, and mechanical suffocation.[8] Figure 1–4 gives a complete breakdown of the percentages for the various categories of causes.

WORK INJURIES BY TYPE OF ACCIDENT

Work injuries can be classified by the type of accident from which they resulted. The most common causes of work injuries are

- Overexertion
- Impact accidents
- Falls
- Bodily reaction (to chemicals)
- Compression
- Motor vehicle accidents
- Exposure to radiation/caustics
- Rubbing or abrasions
- Exposure to extreme temperatures

Overexertion, the result of employees working beyond their physical limits, is the leading cause of work injuries. According to the National Safety Council, almost 31 percent of all work injuries are caused by overexertion. Impact accidents involve a worker being struck by or against an object. The next most prominent cause of work injuries is falls.[9] The remaining accidents are distributed fairly equally among the other causes just listed.

PARTS OF THE BODY INJURED ON THE JOB

In order to develop and maintain an effective safety and health program, it is necessary to know not only the most common causes of death and injury but also the parts of the body most frequently injured.

Typically, the most frequent injuries to specific parts of the body are as follows (from most frequent to least):

1. Back
2. Legs and fingers
3. Arms and multiple parts of the body
4. Trunk
5. Hands

Figure 1–4
Work deaths by cause for a typical year.

Cause	Percent
Motor vehicle related	37.2%
Falls	12.5
Electric current	3.7
Drowning	3.2
Fire related	3.1
Air transport related	3.0
Poison (solid, liquid)	2.7
Water transport related	1.6
Poison (gas, vapor)	1.4
Other	31.6

6. Eyes, head, and feet
7. Neck, toes, and body systems

ESTIMATING THE COST OF ACCIDENTS

Even decision makers who support accident prevention must consider the relative costs of such efforts. To do this, they must be able to estimate the cost of accidents. The procedure for estimating costs set forth in this section was developed by Professor Rollin H. Simonds of Michigan State College working in conjunction with the Statistics Division of the National Safety Council.

Cost Estimation Method

Professor Simonds states that in order to have value, a cost estimate must relate directly to the specific company in question. Applying broad industry cost factors will not suffice. To arrive at company-specific figures, Simonds recommends that costs associated with an accident be divided into *insured* and *uninsured* costs.[10]

Determining the insured costs of accidents is a simple matter of examining accounting records. The next step involves calculating the uninsured costs. Simonds recommends that accidents be divided into the following four classes:

- *Class 1 accidents.* Lost workdays, permanent partial disabilities, and temporary total disabilities.
- *Class 2 accidents.* Treatment by a physician outside of the company's facility.
- *Class 3 accidents.* Locally provided first aid, property damage of less than $100, or the loss of fewer than eight hours of work time.
- *Class 4 accidents.* Injuries that are so minor they do not require the attention of a physician, result in property damage of $100 or more, or cause eight or more work hours to be lost.[11]

Average uninsured costs for each class of accident can be determined by pulling the records of all accidents that occurred during a specified period and sorting the records according to class. For each accident in each class, record every cost that was not covered by insurance. Compute the total of these costs by class of accident and divide by the total number of accidents in that class to determine an average uninsured cost for each class, specific to the particular company.

Figure 1–5 is an example of how the average cost of a selected sample of Class 1 accidents might be determined. In this example, there were four Class 1 accidents in the pilot test. These four accidents cost the company a total of $554.23 in uninsured costs, or an average of $138.56 per accident. Using this information, accurate estimates of the cost of an accident can be made as can accurate predictions.

Other Cost Estimation Methods

The costs associated with workplace accidents, injuries, and incidents fall into broad categories such as the following:

- Lost work hours
- Medical costs
- Insurance premiums and administration
- Property damage
- Fire losses
- Indirect costs

Class of Accident	Accident Number							
Class 1	1	2	3	4	5	6	7	8
Cost A	16.00	6.95	15.17	3.26				
Cost B	72.00	103.15	97.06	51.52				
Cost C	26.73	12.62	—	36.94				
Cost D	—	51.36	—	38.76				
Cost E	—	11.17	—	24.95				
Cost F	—	—	—	−13.41				
Cost G	—	—	—	—				
Total	114.73	185.25	112.23	142.02				

Grand Total: $554.23

Average Cost per Accident: $138.56 (grand total ÷ number of accidents)

Signature: _____ Date: _____

Figure 1–5
Uninsured costs worksheet.

Calculating the direct costs associated with lost work hours involves compiling the total number of lost hours for the period in question and multiplying the hours times the applicable loaded labor rate. The loaded labor rate is the employee's hourly rate plus benefits. Benefits vary from company to company but typically inflate the hourly wage by 20 to 35 percent. A sample cost-of-lost-hours computation follows:

$$\text{Employee Hours Lost (4th quarter)} \times \text{Average Loaded Labor Rate} = \text{Cost}$$
$$386 \times \$13.48 = \$5{,}203.28$$

In this example, the company lost 386 hours due to accidents on the job in the fourth quarter of its fiscal year. The employees who actually missed time at work formed a pool of people with an average loaded labor rate of $13.48 per hour ($10.78 average hourly wage plus 20 percent for benefits). The average loaded labor rate multiplied times the 386 lost hours reveals an unproductive cost of $5,203.28 to this company.

By studying records that are readily available in the company, a safety professional can also determine medical costs, insurance premiums, property damage, and fire losses for the time period in question. All of these costs taken together result in a subtotal cost. This figure is then increased by a standard percentage to cover indirect costs to determine the total cost of accidents for a specific time period. The percentage used to calculate indirect costs can vary from company to company, but 20 percent is a widely used figure.

══════════ **APPLICATION SCENARIOS** ══════════

1. The CEO of your company wants a monthly report of accidents within the company and has asked you to draft a format for him. The report is to contain the number of accidents per month by category, but the CEO did not specify the categories. The report is also to contain the costs of accidents by categories (again, unspecified). Develop a report format.

2. You have just been hired to help build a small start-up company into a full-fledged operation with more than 100 employees in the processing plant and 25 more employees in the office. One of your first tasks is to develop a prediction of the

types of injuries that are likely to occur once the plant is operational. Your list will be used to develop an accident-prevention program. What types of injuries would you expect in such an operation?

3. You have been asked to develop a method for estimating the cost of accidents at your company. Develop the method and include clear instructions so that it can be easily used by your company or another company.

ENDNOTES

1. *Accident Facts,* 1991 ed. (Chicago: National Safety Council), p. 37.
2. Ibid., p. 25.
3. Business Publishers, Inc. *Occupational Health & Safety Letter,* April 17, 1991, Vol. 21, No. 8, p. 66.
4. *Accident Facts,* 1994–1997 eds. (Chicago: National Safety Council), pp. 4–5.
5. Ibid., p. 34.
6. Ibid., p. 35.
7. Ibid., p. 35.
8. Ibid., p. 36.
9. Ibid., p. 36.
10. *Accident Prevention Manual for Industrial Operations: Administration and Programs,* 9th ed. (Chicago: National Safety Council, 1988), p. 158.
11. Ibid., p. 158.

All About OSHA

Since the early 1970s, the amount of legislation passed—and the number of subsequent regulations—concerning workplace safety and health have increased markedly. Of all the legislation, by far the most significant has been the Occupational Safety and Health Act of 1970, called here the OSHAct.

OSHA'S MISSION AND PURPOSE

According to the Department of Labor, OSHA's mission and purpose can be summarized as follows:

- Encourage employers and employees to reduce workplace hazards.
- Implement new safety and health programs.
- Improve existing safety and health programs.
- Encourage research that will lead to innovative ways of dealing with workplace safety and health problems.
- Establish the rights of employers regarding the improvement of workplace safety and health.
- Establish the rights of employees regarding the improvement of workplace safety and health.

Checklist of Exempted Employers

✓ Persons who are self-employed

✓ Family farms that employ only immediate members of the family

✓ Federal agencies covered by other federal statutes (in cases where these other federal statutes do not cover working conditions in a specific area or areas, OSHA standards apply)

✓ State and local governments (except to gain OSHA's approval of a state-level safety and health plan, states must provide a program for state and local government employees that is at least equal to its private sector plan)

✓ Coal mines (coal mines are regulated by mining-specific laws)

Figure 2–1
Checklist of exempted employees.

■ Monitor job-related illnesses and injuries through a system of reporting and record keeping.

■ Establish training programs to increase the number of safety and health professionals and to improve their competence continually.

■ Establish mandatory workplace safety and health standards and enforce those standards.

■ Provide for the development and approval of state-level workplace safety and health programs.

■ Monitor, analyze, and evaluate state-level safety and health programs.[1]

OSHAct COVERAGE

The OSHAct applies to most employers. If an organization has even one employee, it is considered an employer and must comply with applicable sections of the act. This includes all types of employers from manufacturing and construction to retail and service organizations. There is no exemption for small businesses, although organizations with ten or fewer employees are exempted from OSHA inspections and the requirement to maintain injury/illness records.

Although the OSHAct is the most comprehensive and far-reaching piece of safety and health legislation ever passed in this country, it does not cover all employers. In general, the OSHAct covers employers in all 50 states, the District of Columbia, Puerto Rico,

Figure 2–2
Checklist of OSHA requirements
(areas of concern).

**Checklist of OSHA Requirements
(Areas of Concern)**

✓ Fire protection

✓ Electricity

✓ Sanitation

✓ Air quality

✓ Machine use, maintenance, and repair

✓ Posting of notices and warnings

✓ Reporting of accidents and illnesses

✓ Maintaining written compliance programs

✓ Employee training

---

and all other territories that fall under the jurisdiction of the U.S. government. Exempted employers are summarized in the checklist in Figure 2–1.

Federal government agencies are required to adhere to safety and health standards that are comparable to and consistent with OSHA standards for private sector employees. OSHA evaluates the safety and health programs of federal agencies. However, OSHA cannot assess fines or monetary damages against other federal agencies as it can against private sector employers.

There are many OSHA requirements to which employers must adhere. Some apply to all employers—except those exempted—whereas others apply only to specific types of employers. These requirements cover such areas of concern as those summarized in Figure 2–2.

OSHA STANDARDS

The following statement by the U.S. Department of Commerce summarizes OSHA's responsibilities relating to standards:

> In carrying out its duties, OSHA is responsible for promulgating legally enforceable standards. OSHA standards may require conditions, or the adoption or use of one or more practices, means, methods, or processes reasonably necessary and appropriate to protect workers on the job. It is the responsibility of employers to become familiar with standards applicable to their establishments and to ensure that employees have and use personal protective equipment when required for safety.[2]

The general duty clause of the OSHAct requires that employers provide a workplace that is free from hazards that are likely to harm employees. This is important because the general duty clause applies when there is no specific OSHA standard for a given situation. Where OSHA standards do exist, employers are required to comply with them as written.

How to Read an OSHA Standard

OSHA standards are typically long and complex and are written in the language of lawyers and bureaucrats, making them difficult to read. However, reading OSHA standards can be simplified somewhat if one understands the system.

OSHA standards are part of the Code of Federal Regulations (C.F.R.) published by the Office of the Federal Register. The regulations of all federal-government agencies are published in the C.F.R. Title 29 contains all of the standards assigned to OSHA. Title 29 is divided into several parts, each carrying a four number designator (such as Part 1901, Part 1910). These parts are divided into sections, each carrying a numerical designation. For example, 29 C.F.R. 1910.1 means *Title 29, Part 1910, Section 1, Code of Federal Regulations*.

The sections are divided into four different levels of subsections, each with a particular type of designator as follows:

First Level:	Alphabetically using lowercase letters in parentheses: (a) (b) (c) (d)
Second Level:	Numerically using numerals in parentheses: (1) (2) (3) (4)
Third Level:	Numerically using roman numerals: (i) (ii) (iii) (iv)
Fourth Level:	Alphabetically using uppercase letters in parentheses: (A), (B), (C), (D)

Occasionally, the standards go beyond the fourth level of subsection. In these cases, the sequence just described is repeated with the designator shown in parentheses underlined. For example: (a), (1), (i), (A).

Understanding the system used for designating sections and subsections of OSHA standards can guide readers more quickly to the specific information needed. This helps to reduce the amount of cumbersome reading needed to comply with the standards.

Temporary Emergency Standards

The procedures described in the previous section apply in all cases. However, OSHA is empowered to pass temporary emergency standards on an emergency basis without undergoing normal adoption procedures. Such standards remain in effect only until permanent standards can be developed.

To justify passing temporary standards on an emergency basis, OSHA must determine that workers are in imminent danger from exposure to a hazard not covered by existing standards. Once a temporary standard has been developed, it is published in the *Federal Register*. This step serves as the notification step in the permanent adoption process. At this point, the standard is subjected to all of the other adoption steps outlined in the preceding section.

How to Appeal a Standard

After a standard has been passed, it becomes effective on the date prescribed. This is not necessarily the final step in the appeals process, however. A standard, either permanent or temporary, may be appealed by any person who is opposed to it.

An appeal must be filed with the U.S. Court of Appeals serving the geographic region in which the complainant lives or does business. Appeal paperwork must be initiated within 60 days of a standard's approval. However, the filing of one or more appeals does not delay the enforcement of a standard unless the court of appeals handling the matter mandates a delay. Typically, the new standard is enforced as passed until a ruling on the appeal is handed down.

Requesting a Variance

Occasionally, an employer may be unable to comply with a new standard by the effective date of enforcement. In such cases, the employer may petition OSHA at the state or federal level for a variance. Following are the different types of variances that can be granted.

Temporary Variance

When an employer advises that it is unable to comply with a new standard but may be able to if given additional time, a temporary variance may be requested. OSHA may grant such a variance for up to a maximum of one year. To be granted a temporary variance, employers must demonstrate that they are making a concerted effort to comply and taking the steps necessary to protect employees while working toward compliance.

Application procedures are very specific. Prominent among the requirements are the following: (1) identification of the parts of the standard that cannot be complied with; (2) explanation of the reasons why compliance is not possible; (3) detailed explanations of the steps that have been taken so far to comply with the standard; and (4) explanation of the steps that will be taken to fully comply.

Variances are not granted simply because an employer cannot afford to comply. For example, if a new standard requires employers to hire a particular type of specialist but there is a shortage of people with the requisite qualifications, a temporary variance might be granted. However, if the employer simply cannot afford to hire such a specialist, the variance will probably be denied. Once a temporary variance is granted, it may be renewed twice. The maximum period of each extension is six months.

Permanent Variance

Employers who feel they already provide a workplace that exceeds the requirements of a new standard may request a permanent variance. They present their evidence, which is inspected by OSHA. Employees must be informed of the application for a variance and notified of their right to request a hearing. Having reviewed the evidence and heard testimony (if a hearing has been held), OSHA can award or deny the variance. If a permanent variance is awarded, it comes with a detailed explanation of the employer's on-going responsibilities regarding the variance. If, at any time, the company does not meet these responsibilities, the variance can be revoked.

Other Variances

In addition to temporary and permanent variances, an experimental variance may be awarded to companies that participate in OSHA-sponsored experiments to test the effectiveness of new health and safety procedures. Variances also may be awarded in cases where the secretary of labor determines that a variance is in the best interest of the country's national defense.

When applying for a variance, employers are required to comply with the standard until a decision has been made. If this is a problem, the employer may petition OSHA for an interim order. If granted, the employer is released from the obligation to comply until a decision is made. In such cases, employees must be informed of the order.

Typical of OSHA standards are the confined space and hazardous waste standards. Brief profiles of these standards provide an instructive look at how OSHA standards are structured and the extent of their coverage.

RECORD KEEPING AND REPORTING

One of the breakthroughs of the OSHAct was the centralization and systematization of record keeping. This has simplified the process of collecting health and safety statistics for the purpose of monitoring problems and taking the appropriate steps to solve them.

In January 1997, OSHA made substantial changes to its record keeping and reporting requirements.

Reporting Requirements[3]

All occupational illnesses and injuries must be reported, if they result in one or more of the following:

- Death of one or more workers
- One or more days away from work
- Restricted motion or restrictions to the work that an employee can do
- Loss of consciousness of one or more workers
- Transfer of an employee to another job
- Medical treatment beyond in-house first aid (if it is not on the first-aid list, it is considered medical treatment)
- Any other condition listed in Appendix B of the rule.

Record-Keeping Requirements[4]

Employers are required to keep injury and illness records for each location where they do business. For example, an automobile manufacturer with plants in several states must keep records at each individual plant for that plant. Records must be maintained on an annual basis using special forms prescribed by OSHA. Computer or electronic copies can replace paper copies. Records are not sent to OSHA. Rather, they must be maintained locally for a minimum of three years. However, they must be available for inspection by OSHA at any time.

All records required by OSHA must be maintained on only the following forms:

Establishment Name _____

Establishment Name _____

Mailing Address if Different _____

Industry Description and Standard Industrial Classification (SIC) if known _____

Calendar Year _____

Page _____ of _____

	CASE IDENTIFICATION				CASE DESCRIPTION		CASE CLASSIFICATION			
A. Employee's Name (e.g., Doe, Jane)	**B.** Case #	**C.** Date of Injury or Illness (m/d)	**D.** Department	**E.** Job Title	**F.** Description of Injury or Illness		**G.** Death	**H.** Involving Days Away	**I.** Without Days Away Restricted Work Activity	**Other**	**J.** Employer Use (OTHER)
							(X)	(X) (X Days)	(X)	(X)	
							☐	☐	☐	☐	
							☐	☐	☐	☐	
							☐	☐	☐	☐	
							☐	☐	☐	☐	
							☐	☐	☐	☐	
							☐	☐	☐	☐	
							☐	☐	☐	☐	

Year End Totals _____

Annual Average Number of Employees _____

Total Hours Worked by All Employees _____

I have examined this Log and Summary and certify its accuracy and completeness. X _____ Title _____ Phone (____) _____ Date _____

Figure 2-3

OSHA Injury and Illness Log and Summary based on OSHA Form 300.

14

■ *Log and Summary of Occupational Injuries and Illnesses (OSHA Form 300).* All recordable injuries and illnesses must be recorded on this form (see Figure 2-3) within six working days of when the employer first becomes aware of the situation. To accommodate the use of computers and word processing systems, employers may create a computer version of Form 300 as long as it is clearly a detailed facsimile. Totals recorded must be posted wherever employee notices are usually posted.

■ *Supplementary Record of Occupational Injuries and Illnesses (OSHA Form 301).* Form 300 is a plant-wide summary; Form 301 is a more detailed form for each

---

Employee

1. Last Name _____ First Name _____ MI___
2. Male ☐ Female ☐
3. Date of Birth __/__/__
4. Home Address
5. Date hired __/__/__

Health Care Provider

6. Name of health care provider _____
7. If treatment off-site, facility name and address _____
8. Hospitalized overnight as in-patient?
   (If emergency room only, mark *No*) Yes ☐ No ☐

Illness or Injury

9. Specific injury or illness
   (e.g., Second degree burn or Toxic hepatitis) _____
10. Body part(s) affected (e.g., Lower right forearm) _____
11. Date of injury or illness: __/__/__
12. If employee died, date of death: __/__/__
13. If the case involved days away from work or restricted work activity, enter the date the employee returned to work at full capacity: __/__/__
14. Time of Event: 15. Time employee began work:
    (Specify a.m. or p.m.) _____   (Specify a.m. or p.m.) _____
16. All equipment, materials, or chemicals employee was using when the event occurred. (e.g., Acetylene cutting torch, metal plate)
    _____
    _____

17. Specify activity the employee was engaged in when the event occurred (e.g., Cutting metal plate for flooring). Indicate if activity was part of normal duties.
    _____
    _____
    _____

18. How injury or illness occurred. Describe the sequence of events and include any objects or substances that directly injured or made the employee ill. (e.g., Worker stepped back to inspect work and slipped on some scrap metal. As she fell, worker brushed against the hot metal.)
    _____
    _____
    _____

Employer Use
    _____
    _____
    _____
    _____
    _____

Figure 2-4
OSHA Injury and Illness Record based on OSHA Form 301.

individual injury (see Figure 2-4). Workers' compensation forms or other forms that contain sufficient detail may be substituted for Form 301.

■ *Annual Survey (OSHA Form 200S)*.   This is a special form provided only to employees selected as participants in OSHA's annual statistical survey. Shortly after the end of the year, employers with more than 11 workers receive Form 200S for reporting on the immediate past year. Employers with 11 or fewer workers receive their form at the beginning of the year to be reported on.

Reporting/Record-Keeping Summary

Reporting and record-keeping requirements appear as part of several different OSHA standards. Not all of them apply in all cases. Following is a summary of the most widely applicable OSHA reporting and record-keeping requirements.

■ *29 C.F.R. 1903.2(a) OSHA Poster*.   OSHA Poster 2203, which advises employees of the various provisions of the OSHAct, must be conspicuously posted in all facilities subject to OSHA regulations.

■ *29 C.F.R. 1903.16(a) Posting of OSHA Citations*.   Citations issued by OSHA must be clearly posted for the information of employees in a location as close as possible to the site of the violation.

■ *29 C.F.R. 1904.2 Injury/Illness Log*.   Employers are required to maintain a log and summary of all *recordable* (Form 300) injuries and illnesses of their employees.

■ *29 C.F.R. 1904.4 Supplementary Records*.   Employers are required to maintain supplementary records (OSHA Form 301) that give more complete details relating to all recordable injuries and illnesses of their employees.

■ *29 C.F.R. 1904.5 Annual Summary*.   Employers must complete and post an annual summary of all recordable illnesses and injuries. The summary must be posted from February 1 to March 1 every year.

■ *29 C.F.R. 1904.6 Lifetime of Records*.   Employers are required to keep injury/illness records on file for three years.

■ *29 C.F.R. 1904.7 Access to Records*.   Employers are required to give employees, government representatives, and former employees and their designated representatives access to their own individual injury/illness records.

■ *29 C.F.R. 1904.8 Major Incident Report*.   A major incident is the death of one employee or the hospitalization of five or more employees in one incident. All such incidents must be reported to OSHA.

■ *29 C.F.R. 1904.11 Change of Ownership*.   When a business changes ownership, the new owner is required to maintain the OSHA-related records of the previous owner.

■ *29 C.F.R. 1910.20(d) Medical/Exposure Records*.   Employers are required to maintain medical and/or exposure records for the duration of employment plus 30 years (unless the requirement is superseded by another OSHA standard).

■ *29 C.F.R. 1910.20(e) Access to Medical/Exposure Records*.   Employers that keep medical and/or exposure records are required to give employees access to their own individual records.

■ *29 C.F.R. 1910(g)(1) Toxic Exposure*.   Employees who will be exposed to toxic substances or other harmful agents in the course of their work must be notified when first hired and reminded continually on at least an annual basis thereafter of their right to access to their own individual medical records.

■ *29 C.F.R. 1910.20(g)(2) Distribution of Materials*.   Employers are required to make a copy of OSHA's *Records Access Standard* (29 C.F.R. 1910.20) available to employees. They are also required to distribute to employees any informational materials provided by OSHA.

OSHA makes provisions for awarding record-keeping variances to companies that wish to establish their own record-keeping systems (see variance regulation 1905). Application procedures are similar to those described earlier in this chapter for standard variances. To be awarded a variance, employers must show that their record-keeping system meets or exceeds OSHA's requirements.

Incidence Rates

Two concepts can be important when completing OSHA 200/300 forms: *incidence rates* and *severity rates*. On occasion, it is necessary to calculate the total injury/illness incident rate of an organization in order to complete an OSHA Form 200/301. This calculation must include fatalities and all injuries requiring medical treatment beyond mere first aid.

The formula for determining the total injury/illness incident rate is as follows:

$$IR = N \times 200,000 \div T$$

IR = Total injury/illness incidence rate
N = Number of injuries, illnesses, and fatalities
T = Total hours worked by all employees during the period in question

The number 200,000 in the formula represents the number of hours that 100 employees work in a year (40 hours per week x weeks = 2,000 hours per year per employee). Using the same basic formula with only minor substitutions, safety managers can calculate the following types of incidence rates:

1. Injury rate
2. Illness rate
3. Fatality rate
4. Lost workday cases rate
5. Number of lost workdays rate
6. Specific hazard rate
7. Lost workday injuries rate

The *number of lost workdays rate,* which does not include holidays, weekends, or any other days that employees would not have worked anyway, takes the place of the old *severity rate* calculation.

Record Keeping and Reporting Exceptions

Among the exceptions to OSHA's record-keeping and reporting requirements, the two most prominent are as follows:

- Employers with ten or fewer employees (full or part-time in any combination)
- Employers in one of the following categories: real estate, finance, retail trade, or insurance

There are also partial exceptions to OSHA's record-keeping and reporting requirements. Most businesses that fall into Standard Industrial Classifications (SIC) codes 52–89 are exempt from all record-keeping and reporting requirements except in the case of a fatality or an incident in which five or more employees are hospitalized. Within this range of SIC codes, exemptions do *not* apply to the following classifications:

52:	Building materials/garden supplies
53/54:	General merchandise food stores
70:	Hotels/lodging establishments
75/76:	Repair services

79:        Amusement/recreation services

80:        Health services

KEEPING EMPLOYEES INFORMED

One of the most important requirements of the OSHAct is *communication*. Employers are required to keep employees informed about safety and health issues that concern them. Most of OSHA's requirements in this area concern the posting of material. Employers are required to post the following material at locations where employee information is normally displayed:

- OSHA Poster 2203, which explains employee rights and responsibilities as prescribed in the OSHAct. The state version of this poster may be used as a substitute.
- Summaries of variance requests of all types.
- Copies of all OSHA citations received for failure to meet standards. Unlike other informational material, citations must be posted near the site of the violation. They must remain until the violation is corrected or for a minimum of three days, whichever period is longer.
- The summary page of OSHA Form 200/300 (Log and Summary of Occupational Injuries and Illnesses). Each year the new summary page must be posted by February 1 and must remain posted until March 1.

In addition to the posting requirements, employers must also provide employees who request them with copies of the OSHAct and any OSHA rules that may concern them. Employees must be given access to records of exposure to hazardous materials and medical surveillance that has been conducted.

WORKPLACE INSPECTIONS

One of the methods OSHA uses for enforcing its rules is the workplace inspection. OSHA personnel may conduct workplace inspections unannounced, and except under special circumstances, giving an employer prior notice is a crime punishable by fine, imprisonment, or both.

When OSHA compliance officers arrive to conduct an inspection, they are required to present their credentials to the person in charge. Having done so, they are authorized to enter, at reasonable times, any site, location, or facility where work is taking place. They may inspect, at reasonable times, any condition, facility, machine, equipment, materials, and so on. Finally, they may question, in private, any employee or other person formally associated with the company.

Under special circumstances, employers may be given up to a maximum of 24 hours' notice of an inspection. These circumstances are summarized in the checklist in Figure 2–5.

Employers may require that OSHA have a judicially authorized warrant before conducting an inspection. However, having obtained a legal warrant, OSHA personnel must be allowed to proceed without interference or impediment.

The OSHAct applies to approximately six million work sites in the United States. Sheer volume dictates that OSHA establish priorities for conducting inspections. These priorities are as follows: imminent danger situations, catastrophic fatal accidents, employee complaints, planned high-hazard inspections, and follow-up inspections.

After being scheduled, the inspection proceeds in the following steps:

- The OSHA compliance officer presents his or her credentials to a company official.
- The compliance officer conducts an opening conference with pertinent company officials and employee representatives. The following information is explained during

---

Checklist of Special Circumstances (for Notification of an Inspection)

✓ When imminent danger conditions exist

✓ When special preparation on the part of the employer is required

✓ When inspection must take place at times other than during regular business
   hours

✓ When it is necessary to ensure that the employer, employee representative,
   and other pertinent personnel will be present

✓ When the local area director for OSHA advises that advance notice will result
   in a more effective inspection

Figure 2–5
Checklist of special circumstances for notification of an inspection.

the conference: why the plant was selected for inspection, the purpose of the inspection, its scope, and applicable standards.

- After choosing the route and duration, the compliance officer makes the inspection tour. During the tour, the compliance officer may observe, interview pertinent personnel, examine records, take readings, and make photographs.

- The compliance officer holds a closing conference, which involves open discussion between the officer and company/employee representatives. OSHA personnel advise company representatives of problems noted, actions planned as a result, and assistance available from OSHA.

CITATIONS AND PENALTIES

Based on the findings of the compliance officer's workplace inspections, OSHA is empowered to issue citations and/or assess penalties. A citation informs the employer of OSHA violations. Penalties are typically fines assessed as the result of citations. The types of citations and their corresponding penalties, as quoted from OSHA 2056, 1991 (Revised), are as follows:

- *Other-than-serious violation.*   A violation that has a direct relationship to job safety and health, but probably would not cause death or serious physical harm. A penalty for an other-than-serious violation may be adjusted downward by as much as 95 percent, depending on the employer's good faith (demonstrated efforts to comply with the act), history of previous violations, and size of business.

- *Willful violation.*   A violation that the employer intentionally and knowingly commits. The employer either knows that what he or she is doing constitutes a violation or is aware that a hazardous condition exists and has made no reasonable effort to eliminate it. A proposed penalty for a willful violation may be adjusted downward, depending on the size of the business and its history of previous violations. Usually, no credit is given for good faith. If an employer is convicted of a willful violation of a standard that has resulted in the death of an employee, the offense is punishable by a court-imposed fine or by imprisonment for up to six months, or both.

- *Repeat violation.*   A violation of any standard, regulation, rule, or order where, upon reinspection, a substantially similar violation is found. To be the basis of a repeat citation, the original citation must be final; a citation under contest may not serve as the basis for a subsequent repeat citation.

- *Failure to correct prior violation.*   Failure to correct a prior violation may bring a civil penalty of up to $7,000 for each day that the violation continues beyond the prescribed abatement date.[5]

In addition to the citations and penalties described in the preceding paragraphs, employers may also be penalized by additional fines and/or prison if convicted of any of the following offenses: (1) falsifying records or any other information given to OSHA personnel; (2) failing to comply with posting requirements; and (3) interfering in any way with OSHA compliance officers in the performance of their duties.

THE APPEALS PROCESS

Employee Appeals

Employees may not contest the fact that citations were or were not awarded or the amounts of the penalties assessed. However, they may appeal the following aspects of OSHA's decisions regarding their workplace: (1) the amount of time (abatement period) given an employer to correct a hazardous condition that has been cited, and (2) an employer's request for an extension of an abatement period. Such appeals must be filed within ten working days of a posting. Although opportunities for formal appeals by employees are unlimited, employees may request an informal conference with OSHA officials to discuss any issue relating to the findings and results of a workplace inspection.

Employer Appeals

Employers may appeal a citation, an abatement period, or the amount of a proposed penalty. Before actually filing an appeal, however, an employer may ask for an informal meeting with OSHA's area director. The area director is empowered to revise citations, abatement periods, and penalties in order to settle disputed claims. If the situation is not resolved through this step, an employer may formalize the appeal. Formal appeals are of two types: (1) a petition for modification of abatement, or (2) a notice of contest. The specifics of both are explained in the following paragraphs.

Petition for Modification of Abatement (PMA)

The PMA is available to employers who intend to correct the situation for which a citation was issued, but who need more time. As a first step, the employer must make a good-faith effort to correct the problem within the prescribed timeframe. Having done so, the employer may file a petition for modification of abatement. The petition must contain the following information:

- Descriptions of steps taken so far to comply
- How much additional time is needed for compliance and why
- Descriptions of the steps being taken to protect employees during the interim
- Verification that the PMA has been posted for employee information and that the employee representative has been given a copy

Notice of Contest

An employer who does not wish to comply may contest a citation, an abatement period, and/or a penalty. The first step is to notify OSHA's area director in writing. This is known as filing a notice of contest. It must be done within 15 working days of receipt of a citation or penalty notice. The notice of contest must clearly describe the basis for the employer's challenge and contain all of the information about what is being challenged (i.e., amount of proposed penalty or abatement period, and so on).

Once OSHA receives a notice of contest, the area director forwards it and all pertinent materials to the Occupational Safety and Health Review Commission (OSHRC). OSHRC is an independent agency that is associated with neither OSHA nor the Department of Labor. The Department of Labor describes how OSHRC handles an employer's claim:

The commission assigns the case to an administrative law judge. The judge may disallow the contest if it is found to be legally invalid, or a hearing may be scheduled for a public place near the employer's workplace. The employer and the employees have the right to participate in the hearing; the OSHRC does not require that they be represented by attorneys. Once the administrative law judge has ruled, any party to the case may request further review by OSHRC. Any of the three OSHRC commissioners also may, at his or her own motion, bring a case before the Commission for review. Commission rulings may be appealed to the appropriate U.S. Court of Appeals.[6]

SERVICES AVAILABLE FROM OSHA

In addition to setting standards and inspecting for compliance, OSHA provides services to help employers meet the latest safety and health standards. The services, typically offered at no cost, are intended for smaller companies, particularly those with especially hazardous processes and/or materials. Three categories of services are available from OSHA: consultation, voluntary protection programs, and training and education services.

Consultation Services

Consultation services provided by OSHA include assistance in (1) identifying hazardous conditions; (2) correcting identified hazards; and (3) developing and implementing programs to prevent injuries and illnesses. To arrange consultation services, employers contact the consultation provider in their state (see Figure 2–6).

The actual services are provided by professional safety and health consultants, who are not OSHA employees. They typically work for state agencies or universities and provide consultation services on a contract basis; OSHA provides the funding. OSHA publication 3047, entitled *Consultation Services for the Employer,* may be obtained from the nearest OSHA office.

Voluntary Protection Programs

OSHA's Voluntary Protection Programs (VPPs) serve the following three basic purposes:

- To recognize companies that have incorporated safety and health programs into their overall management system
- To motivate companies to incorporate health and safety programs into their overall management system
- To promote positive, cooperative relationships among employers, employees, and OSHA

OSHA currently operates three programs under the VPP umbrella. These programs are discussed in the following paragraphs.

Star Program

The Star Program recognizes companies that have incorporated safety and health into their regular management system so successfully that their injury rates are below the national average for their industry. This is OSHA's most strenuous program. To be part of the Star Program, a company must demonstrate

- Management commitment
- Employee participation
- An excellent worksite analysis program
- A hazard prevention and control program
- A comprehensive safety and health training program[7]

State	Telephone	State	Telephone
Alabama	205-348-3033	Nebraska	401-471-4717
Alaska	907-264-2599	Nevada	701-789-0546
Arizona	602-255-5795	New Hampshire	603-271-3170
Arkansas	501-682-4522	New Jersey	609-984-3517
California	415-557-2870	New Mexico	505-827-2885
Colorado	303-491-6151	New York	518-457-5468
Connecticut	203-566-4550	North Carolina	919-733-3949
Delaware	302-571-3908	North Dakota	701-224-2348
District of Columbia	202-576-6339	Ohio	614-644-2631
Florida	904-488-3044	Oklahoma	405-528-1500
Georgia	404-894-8274	Oregon	503-378-3272
Guam	9-011 671-646-9246	Pennsylvania	800-381-1241 (Toll-free)
Hawaii	808-548-7510		412-357-2561
Idaho	208-385-3283	Puerto Rico	809-754-2134-2171
Illinois	312-917-2339	Rhode Island	401-277-2438
Indiana	317-232-2688	South Carolina	803-734-9579
Iowa	515-281-5352	South Dakota	605-688-4101
Kansas	913-296-4386	Tennessee	615-741-7036
Kentucky	502-564-6895	Texas	512-458-7254
Louisiana	504-342-9601	Utah	801-530-6868
Maine	207-289-6460	Vermont	801-828-2765
Maryland	301-333-4219	Virginia	804-367-1986
Massachusetts	616-727-3463	Virgin Islands	809-772-1315
Michigan	517-335-8250 (Health)	Washington	206-586-0961
	517-322-1814 (Safety)	West Virginia	304-348-7890
Minnesota	612-297-2393	Wisconsin	608-266-8579 (Health)
Mississippi	601-987-3961		414-512-5063 (Safety)
Missouri	314-751-3403	Wyoming	307-777-7786
Montana	406-444-6401		

Figure 2–6
State consultation project directory.

Merit Program

The Merit Program is less strenuous than the Star Program. It is seen as a stepping-stone to recognize companies that have made a good start toward Star Program recognition. OSHA works with such companies to help them take the next step and achieve Star Program recognition.

Demonstration Program

The Department of Labor describes the Demonstration Program as follows: "for companies that provide Star-quality worker protection in industries where certain Star requirements can be changed to include these companies as Star participants."[8]

Companies participating in any of the VPPs are exempt from regular programmed OSHA inspections. However, employee complaints, accidents that result in serious injury, or major chemical releases will be "handled according to routine enforcement procedures."[9]

Training and Education Services

Training and education services available from OSHA take several forms. OSHA operates a training institute in Des Plaines, Illinois, that offers a wide variety of services to safety and health personnel from the public and private sectors. The institute has a full range of facilities including classrooms and laboratories in which it offers more than 60 courses.

To promote training and education in locations other than the institute, OSHA awards grants to nonprofit organizations. Colleges, universities, and other nonprofit organizations apply for funding to cover the costs of providing workshops, seminars, or short courses on safety and health topics currently high on OSHA's list of priorities. Grant funds must be used to plan, develop, and present instruction. Grants are awarded annually and require a match of at least 20 percent of the total grant amount.

EMPLOYER RIGHTS AND RESPONSIBILITIES

OSHA is very specific in delineating the rights and responsibilities of employers regarding safety and health. These rights and responsibilities, as set forth in OSHA publication 2056, are summarized in this section.

Employer Rights

The following list of employer rights under the OSHAct is adapted from OSHA 2056, 1991 (Revised). Employers have the right to do the following:

- Seek advice and consultation as needed by contacting or visiting the nearest OSHA office.
- Request proper identification of the OSHA compliance officer prior to an inspection.
- Be advised by the compliance officer of the reason for an inspection.
- Have an opening and closing conference with the compliance officer in conjunction with an inspection.
- Accompany the compliance officer on the inspection.
- File a notice of contest with the OSHA area director within 15 working days of receipt of a notice of citation and proposed penalty.
- Apply for a temporary variance from a standard if unable to comply because the materials, equipment, or personnel needed to make necessary changes within the required time are not available.
- Apply for a permanent variance from a standard if able to furnish proof that the facilities or methods of operation provide employee protection at least as effective as that required by the standard.
- Take an active role in developing safety and health standards through participation in OSHA Standards Advisory Committees, through nationally recognized standards-setting organizations, and through evidence and views presented in writing or at hearings.
- Be assured of the confidentiality of any trade secrets observed by an OSHA compliance officer during an inspection.
- Ask NIOSH for information concerning whether any substance in the workplace has potentially toxic effects.[10]

Employer Responsibilities[11]

In addition to the rights set forth in the previous subsection, employers have prescribed responsibilities. The following list of employer responsibilities under the OSHAct is adapted from OSHA 2056 (Revised). Employers must do the following:

- Meet the general duty responsibility to provide a workplace free from hazards that are causing or are likely to cause death or serious physical harm to employees, and comply with standards, rules, and regulations issued under the OSHAct.
- Be knowledgeable of mandatory standards and make copies available to employees for review upon request.
- Keep employees informed about OSHA.
- Continually examine workplace conditions to ensure that they conform to standards.
- Minimize or reduce hazards.
- Make sure employees have and use safe tools and equipment (including appropriate personal protective equipment) that is properly maintained.
- Use color codes, posters, labels, or signs as appropriate to warn employees of potential hazards.
- Establish or update operating procedures and communicate them so that employees follow safety and health requirements.
- Provide medical examinations when required by OSHA standards.
- Provide the training required by OSHA standards.
- Report to the nearest OSHA office within 48 hours any fatal accident or one that results in the hospitalization of five or more employees.
- Keep OSHA-required records of injuries and illnesses and post a copy of the totals from the last page of OSHA Form 200/300 during the entire month of February each year. (This applies to employers with 11 or more employees.)
- At a prominent location within the workplace, post OSHA Poster 2203 informing employees of their rights and responsibilities.
- Provide employees, former employees, and their representatives access to the Log and Summary of Occupational Injuries and Illnesses (OSHA Form 200/300) at a reasonable time and in a reasonable manner.
- Give employees access to medical and exposure records.
- Give the OSHA compliance officer the names of authorized employee representatives who may be asked to accompany the compliance officer during an inspection.
- Not discriminate against employees who properly exercise their rights under the act.
- Post OSHA citations at or near the work site involved. Each citation or copy must remain posted until the violation has been abated or for three working days, whichever is longer.
- Abate cited violations within the prescribed period.

EMPLOYEE RIGHTS AND RESPONSIBILITIES

Employee Rights

Section 11(c) of the OSHAct delineates employee rights. These rights are actually protection against punishment for employees who exercise their right to pursue any of the following courses of action:

- Complain to an employer, union, OSHA, or any other government agency about job safety and health hazards.
- File safety or health grievances.
- Participate in a workplace safety and health committee or in union activities concerning job safety and health.
- Participate in OSHA inspections, conferences, hearings, or other OSHA-related activities.[12]

Employees who feel they are being treated unfairly because of actions they have taken in the interest of safety and health have 30 days in which to contact the nearest OSHA office. Upon receipt of a complaint, OSHA conducts an investigation and makes recommendations based on its findings. If an employer refuses to comply, OSHA is empowered to pursue legal remedies at no cost to the employee who filed the original complaint.

In addition to those just set forth, employees have a number of other rights. Employees may

- Expect employers to make review copies available of OSHA standards and requirements.
- Ask employers for information about hazards that may be present in the workplace.
- Ask employers for information on emergency procedures.
- Receive safety and health training.
- Be kept informed about safety and health issues.
- Anonymously ask OSHA to conduct an investigation of hazardous conditions at the work site.
- Be informed of actions taken by OSHA as a result of a complaint.
- Observe during an OSHA inspection and respond to the questions asked by a compliance officer.
- See records of hazardous materials in the workplace.
- See their medical record.
- Review the annual Log and Summary of Occupational Injuries (OSHA Form 200/300).
- Have an exit briefing with the OSHA compliance officer following an OSHA inspection.
- Anonymously ask NIOSH to provide information about toxicity levels of substances used in the workplace.
- Challenge the abatement period given employers to correct hazards discovered in an OSHA inspection.
- Participate in hearings conducted by the Occupational Safety and Health Review Commission.
- Be advised when an employer requests a variance to a citation or any OSHA standard.
- Testify at variance hearings.
- Appeal decisions handed down at OSHA variance hearings.
- Give OSHA input concerning the development, implementation, modification, and/or revocation of standards.[13]

Employee Responsibilities

Employees have a number of specific responsibilities. The following list of employee responsibilities is adapted from OSHA 2056 (Revised). Employees must

- Read the OSHA poster at the job site and be familiar with its contents.
- Comply with all applicable OSHA standards.
- Follow safety and health rules and regulations prescribed by the employer and promptly use personal protective equipment while engaged in work.
- Report hazardous conditions to the supervisor.
- Report any job-related injury or illness to the employer and seek treatment promptly.
- Cooperate with the OSHA compliance officer conducting an inspection.
- Exercise their rights under the OSHAct in a responsible manner.[14]

KEEPING UP TO DATE ON OSHA

OSHA's standards, rules, and regulations are always subject to change. The development, modification, and revocation of standards is an ongoing process. It is important for prospective and practicing safety and health professionals to stay up to date with the latest actions and activities of OSHA. Following is an annotated list of strategies that can be used to keep current:

■ Establish contact with the nearest regional or area OSHA office and periodically request copies of new publications or contact the OSHA Publications Office at the following address:

> OSHA Publications Office
> 200 Constitution Avenue, N.W.
> Room N–3101
> Washington, D.C. 20210

or at OSHA's Internet address:

> http://www.osha.gov

■ Review professional literature in the safety and health field. Numerous periodicals carry OSHA updates that are helpful.

■ Establish and maintain relationships with other safety and health professionals for the purpose of sharing information, and do so frequently.

■ Join professional organizations, review their literature, and attend their conferences.

OSHA'S GENERAL INDUSTRY STANDARDS

The most widely applicable OSHA standards are the *General Industry Standards*. These standards are found in 29 C.F.R. 1910. Part 1910 consists of 21 subparts, each carrying an uppercase-letter designation. Subparts A and B contain no compliance requirements. The remaining subparts are described in the following subsections.

Subpart C: General Safety and Health Provisions

The only compliance standard in Subpart C is *Access to Employee Exposure and Medical Records*. Employers that are required to keep medical and exposure records must do the following: (1) maintain the records for the duration of employment plus 30 years, and (2) give employees access to their individual personal records.

Subpart D: Walking–Working Surfaces

Subpart D contains the standards for all surfaces on which employees walk or work. Specific sections of Subpart D are as follows:

1910.21	Definitions
1910.22	General requirements
1910.23	Guarding floor and wall openings and holes
1910.24	Fixed industrial stairs
1910.25	Portable wood ladders
1910.26	Portable metal ladders
1910.27	Fixed ladders
1910.28	Safety requirements for scaffolding
1910.29	Manually propelled mobile ladder stands and scaffolds (towers)

DISCUSSION CASE

What Is Your Opinion?

"The OSHAct is a nightmare! All it has accomplished is the creation of a department full of governmental bureaucrats who bully private industry." This was the opening line in a debate on government safety regulations sponsored by the Industrial Technology Department of Pomona State University. The OSHA advocate in the debate responded as follows: "The OSHAct is a model of government regulation as it should be. Had private industry been responsive to the safety and health concerns of employees, there would have been no need for government regulation." These are two widely divergent viewpoints. What is your opinion in this matter?

1910.30	Other working surfaces
1910.31	Sources of standards
1910.32	Standards organizations

Subpart E: Means of Egress

Subpart E requires employers to ensure that employees have a safe, accessible, and efficient means of escaping a building under emergency circumstances. Specific sections of Subpart E are as follows:

1910.35	Definitions
1910.36	General requirements
1910.37	Means of egress, general
1910.38	Employee emergency plans and fire prevention plan
1910.39	Source of standards
1910.40	Standards organizations

Emergency circumstances might be caused by fire, explosions, hurricanes, tornadoes, flooding, terrorist acts, earthquakes, nuclear radiation, or other acts of nature not listed here.

Subpart F: Powered Platforms

Subpart F applies to powered platforms, mechanical lifts, and vehicle-mounted work platforms. The requirements of this subpart apply only to employers who use this type of equipment in facility maintenance operations. Specific sections of Subpart F are as follows:

1910.66	Powered platforms for building maintenance
1910.67	Vehicle-mounted elevating and rotating work platforms
1910.68	Manlifts
1910.69	Sources of standards
1910.70	Standards organizations

Subpart G: Health and Environmental Controls

The most widely applicable standard in Subpart G is 1910.95 (occupational noise exposure). Other standards in this subpart pertain to situations where ionizing and/or non-ionizing radiation are present. Specific sections of Subpart G are as follows:

1910.94	Ventilation
1910.95	Occupational noise exposure
1910.96	Ionizing radiation
1910.97	Nonionizing radiation
1910.98	Effective dates
1910.99	Sources of standards
1910.100	Standards organizations

Subpart H: Hazardous Materials

Four of the standards in Subpart H are widely applicable. Section 1901.106 is an extensive standard covering the use, handling, and storage of flammable and combustible liquids. Of particular concern are fire and explosions. Section 1910.107 applies to indoor spray-painting processes and processes in which paint is applied in powder form (e.g., electrostatic powder spray).

Section 1910.119 applies to the management of processes involving specifically named chemicals and flammable liquids and gases. Section 1910.120 contains requirements relating to emergency response operations and hazardous waste. All of the standards contained in Subpart H are as follows:

1910.101	Compressed gases (general requirements)
1910.102	Acetylene
1910.103	Hydrogen
1910.104	Oxygen
1910.105	Nitrous oxide
1910.106	Flammable and combustible liquids
1910.107	Spray finishing using flammable and combustible materials
1910.108	Dip tanks containing flammable and combustible materials
1910.109	Explosive and blasting agents
1910.110	Storage and handling of liquefied petroleum gases
1910.111	Storage and handling of anhydrous ammonia
1910.114	Effective dates
1910.115	Sources of standards
1910.116	Standards organizations
1910.119	Process safety management of highly hazardous chemicals
1910.120	Hazardous waste operations and emergency response

Subpart I: Personal Protective Equipment

Subpart I contains three of the most widely applicable standards: 1910.132 General Requirements; 1910.133 Eye and Face Protection; and 1910.134 Respiratory Protection. The most frequently cited OSHA violations relate to these and the other personal-protective equipment standards. All of the standards in this subpart are as follows:

1910.132	General requirements
1910.133	Eye and face protection
1910.134	Respiratory protection

1910.135	Occupational head protection
1910.136	Occupational foot protection
1910.137	Electrical protective devices
1910.138	Effective dates
1910.139	Sources of standards
1910.140	Standards organizations

Subpart J: General Environment Controls

This subpart contains standards that are widely applicable because they pertain to general housekeeping requirements. An especially important standard contained in this subpart is 1910.146: Permit-Required Confined Spaces. A confined space is one that meets any or all of the following criteria:

■ Large enough and so configured that a person can enter it and perform assigned work tasks therein

■ Continuous employee occupancy is not intended

The *lockout/tagout* standard is also contained in this subpart. All of the standards in this subpart are as follows:

1910.141	Sanitation
1910.142	Temporary labor camps
1910.144	Safety color code for marking physical hazards
1910.145	Accident prevention signs and tags
1910.146	Permit-required confined space
1910.147	Control of hazardous energy (lockout/tagout)
1910.148	Standards organizations
1910.149	Effective dates
1910.150	Sources of standards

Subpart K: Medical and First Aid

This is a short subpart, the most important section of which pertains to eye-flushing. If employees are exposed to *injurious corrosive materials,* equipment must be provided for quickly flushing the eyes and showering the body. The standard also requires medical personnel to be readily available. "Readily available" can mean that there is a clinic or hospital nearby. If such a facility is not located nearby, employers must have a person on hand who has had first-aid training. The standards in this subpart are as follows:

| 1910.151 | Medical seminars and first aid |
| 1910.153 | Sources of standards |

Subpart L: Fire Protection

This subpart contains the bulk of OSHA's fire protection standard. These standards detail the employer's responsibilities concerning fire brigades, portable fire-suppression equipment, fixed fire-suppression equipment, and fire-alarm systems. Employers are not required to form fire brigades, but if they choose to, employers must adhere to the standard set forth in 1910.156. The standards in this subpart are as follows:

Fire Protection

| 1910.155 | Scope, application, and definitions applicable to this subpart |
| 1910.156 | Fire brigades |

Portable Fire-Suppression Equipment

| 1910.157 | Portable fire extinguishers |
| 1910.158 | Standpipe and hose systems |

Fixed Fire-Suppression Equipment

1910.159	Automatic sprinkler systems
1910.160	Fixed extinguishing systems, general
1910.161	Fixed extinguishing systems, dry chemical
1910.162	Fixed extinguishing systems, gaseous agent
1910.163	Fixed extinguishing systems, water spray and foam

Other Fire Protection Systems

| 1910.164 | Fire detection systems |
| 1910.165 | Employee alarm systems |

Subpart M: Compressed Gas/Air

This subpart contains just three sections and only on standard 1910.169. This standard applies to compressed-air equipment that is used in drilling, cleaning, chipping, and hoisting. There are many other uses of compressed air in work settings, but 1910.169 applies only to these applications. The sections in this subpart are as follows:

1910.169	Air receivers
1910.170	Sources of standards
1910.171	Standards organizations

Subpart N: Materials Handling and Storage

This subpart contains one broad standard—1910.176—and several more specific standards relating to 1910.176. Subpart N is actually limited in scope. It applies only to the handling and storage of materials, changing rim wheels on large vehicles, and the proper use of specific equipment identified in the standards' titles. All of the standards in this subpart are as follows:

1910.176	Handling materials—general
1910.177	Servicing multi-piece and single-piece rim wheels
1910.178	Powered industrial trucks
1910.179	Overhead and gantry cranes
1910.180	Crawler locomotive and truck cranes
1910.181	Derricks
1910.182	Effective dates
1910.183	Helicopters

1910.184   Slings
1910.189   Sources of standards
1910.190   Standards organizations

Subpart O: Machinery and Machine Guarding

This subpart contains standards relating to specific types of machines. The types of machines covered are identified in the titles of the standards contained in Subpart O. These standards are as follows:

1910.211   Definitions
1910.212   General requirements for all machines
1910.213   Woodworking machinery requirements
1910.214   Cooperage machinery
1910.215   Abrasive wheel machinery
1910.216   Mills and calendars in the rubber and plastics industries
1910.217   Mechanical power presses
1910.218   Forging machines
1910.219   Mechanical power-transmission apparatus
1910.220   Effective dates
1910.221   Sources of standards
1910.222   Standards organizations

Subpart P: Hand Tools/Portable Power Tools

This subpart contains standards relating to the use of hand tools, portable power tools, and compressed-air-powered tools. The types of tools covered in this subpart, in addition to typical hand tools, include jacks, saws, drills, sanders, grinders, planers, power lawn-mowers, and other tools. The standards contained in this subpart are as follows:

1910.241   Definitions
1910.242   Hand- and portable-powered tools and equipment, general
1910.243   Guarding of portable tools and equipment
1910.244   Other portable tools and equipment
1910.245   Effective dates
1910.246   Sources of standards
1910.247   Standards organizations

Subpart Q: Welding, Cutting, and Brazing

Welding, cutting, and brazing are widely used processes. This subpart contains the standards relating to these processes in all of their various forms. The primary safety and health concerns are fire protection, employee personal protection, and ventilation. The standards contained in this subpart are as follows:

1910.251   Definitions
1910.252   General requirements
1910.253   Oxygen-fuel gas welding and cutting
1910.254   Arc welding and cutting

1910.255	Resistance welding
1910.256	Sources of standards
1910.257	Standards organizations

Subpart R: Special Industries

This subpart is different from others in Part 1910. Whereas other subparts deal with specific processes, machines, and materials, Subpart R deals with specific industries. Each separate standard relates to a different category of industry. The standards contained in this subpart are as follows:

1910.261	Pulp, paper, and paperboard mills
1910.262	Textiles
1910.263	Bakery equipment
1910.264	Laundry machinery and operations
1910.265	Sawmills
1910.266	Pulpwood logging
1910.268	Telecommunications
1910.272	Grain handling facilities
1910.274	Sources of standards
1910.275	Standards organizations

Subpart S: Electrical

This subpart contains standards divided into the following two categories: (1) design of electrical systems, and (2) safety-related work practices. These standards are excerpted directly from the National Electrical Code. Those included in Subpart S are as follows:

1910.301	Introduction
1910.302	Electric utilization systems
1910.303	General requirements
1910.304	Wiring design and protection
1910.305	Wiring methods, components, and equipment for general use
1910.306	Specific-purpose equipment and installations
1910.307	Hazardous (classified) locations
1910.308	Special systems
1910.331	Scope
1910.332	Training
1910.333	Selection and use of work practices
1910.334	Use of equipment
1910.335	Safeguards for personnel protection
1910.399	Definitions applicable to this subpart

Subpart T: Commercial Diving Operations

This subpart applies only to commercial diving enterprises. The standards contained in Subpart T are divided into six categories: (1) general, (2) personnel requirements, (3) general operations and procedures, (4) specific operations and procedures, (5) equipment procedures and requirements, and (6) record keeping. These standards are as follows:

1910.401	Scope and application
1910.402	Definitions
1910.410	Qualifications of dive teams
1910.420	Safe practices manual
1910.421	Pre-dive procedure
1910.422	Procedures during dive
1910.423	Post-dive procedures
1910.424	SCUBA diving
1910.425	Surface-supplied-air diving
1910.426	Mixed-gas diving
1910.427	Liveboating
1910.430	Equipment
1910.440	Record keeping
1910.441	Effective dates

Subpart Z: Toxic and Hazardous Substances

This is an extensive subpart containing the standards that establish *permissible exposure limits* (PELs) for numerous toxic and hazardous substances. All such substances have an assigned PEL which is the amount of a given airborne substance to which employees can be exposed during a specified period of time.

The standards relating to these specific toxic and hazardous substances are contained in 1910.1000 through 1910.1500 and are as follows:

1910.1000	Air contaminants
1910.1001	Asbestos
1910.1002	Coal tar pitch volatiles; interpretation of term
1910.1003	2-Nitrobiphenyl
1910.1004	Alpha-Nephthylamine
1910.1006	Methyl chloromethyl ether
1910.1007	3,3¢–Dichlorobenzidine (and its salts)
1910.1008	Bis-Chloromethyl ether
1910.1009	Beta-Naphthylamine
1910.1010	Benzedrine
1910.1011	2-Aminodiphenyl
1910.1012	Ethyleneimine
1910.1013	Beta-Propiolactone
1910.1014	2-Acetylaminofluorene
1910.1015	2-Dimethylaminoazobenzene
1910.1016	N-Nitrosodimethylamine
1910.1017	Vinyl chloride
1910.1018	Inorganic arsenic
1910.1025	Lead
1910.1027	Cadmium
1910.1028	Benzene
1910.1029	Coke oven emissions

1910.1030	Bloodborne pathogens
1910.1043	Cotton dust
1910.1044	1,2-dibromo-3-chloropropane
1910.1045	Acrylonitrile
1910.1047	Ethylene oxide
1910.1050	Methylenedianiline
1910.1200	Hazard communication
1910.1450	Occupational exposure to hazardous chemicals in laboratories
1910.1499	Sources of standards
1910.1500	Standards organizations

OSHA'S CONSTRUCTION STANDARDS

These standards apply to employers involved in construction, alteration, and/or repair activities. To further identify the scope of the applicability of its construction standards, OSHA took the terms *construction, alteration,* and *repair* directly from the Davis-Bacon Act. This act provides minimum wage protection for employees working on construction projects. The implication is that if the Davis-Bacon Act applies to an employer, OSHA's Construction Standards also apply.

These standards are contained in Part 1926 of the C.F.R. Subparts A–Z. OSHA does not base citations on material contained in Subparts A and B. Consequently, those subparts have no relevance here. The remaining subparts are as follows:

Subpart C	General safety and health provisions
Subpart D	Occupational health and environmental controls
Subpart E	Personal protective and life-saving equipment
Subpart F	Fire protection and prevention
Subpart G	Signs, signals, and barricades
Subpart H	Materials handling, storage, use, and disposal
Subpart I	Tools—hand and power
Subpart J	Welding and cutting
Subpart K	Electrical
Subpart L	Scaffolding
Subpart M	Floor and wall openings
Subpart N	Cranes, derricks, hoists, elevators, and conveyors
Subpart O	Motor vehicles, mechanized equipment, and marine operations
Subpart P	Excavations
Subpart Q	Concrete and masonry construction
Subpart R	Steel erection
Subpart S	Underground construction, caissons, and cofferdams.
Subpart T	Demolition
Subpart U	Blasting and the use of explosives
Subpart V	Power transmission and distribution
Subpart W	Rollover protective structures; overhead protection
Subpart X	Stairways and ladders
Subpart Y	Commercial diving operations
Subpart Z	Toxic and hazardous substances

=========== APPLICATION SCENARIOS ===========

1. A colleague makes the following comment: "All I ever hear is OSHA this and OSHA that. What is the purpose of OSHA anyway?" How would you explain OSHA's purpose to this individual?

2. At a national conference, the CEO of a new small business asks you the following question: "Is my company exempt from OSHA compliance? I have only nine employees." What would you tell her?

3. Having carefully studied the applicable subparts of 29 C.F.R. 1910, you find no regulations relating to a safety issue that has been brought to your attention. Does this mean your company has no responsibility? Should you pursue the issue?

4. Assume that you are strongly opposed to a new OSHA regulation adopted yesterday. What can you do? How should you do it?

5. The deadline for complying with a specific OSHA regulation is next week. Your company will not be ready. What can you do?

6. An employee is slightly injured on the job. He is given first aid treatment and allowed to return to work. This employee is scheduled to work overtime, and he needs the money. Consequently, he wants to keep working. The treatment provided is not on your company's first aid list. What are your reporting requirements?

7. You just came onboard to get a new spin-off company up and running. The company has 123 employees, and products are being shipped. You want to make sure that the company complies with OSHA regulations. What records do the company need to keep? What must be done to keep employees informed concerning OSHA, rights, responsibilities, and so on?

8. An OSHA inspector shows up unannounced at your company. The company's CEO sends the inspector away. Has the CEO violated federal law? Assume that the inspector chooses not to make an issue of being denied access but comes back three days later with a warrant. How does the process go forward at this point?

9. Your company has been fined by OSHA for what the inspector calls a "willful violation." An angry employee comes to you and complains that the fine is not enough. He says, "I've talked to the other employees. This company is being slapped on the wrist when it should really be punished. We want to appeal the fine." What should you tell this employee?

10. As a result of the fine levied in Scenario 9, your company's CEO tells you: "I don't want this problem to ever come up again. OSHA fines add no value to our product, and they detract from the bottom line. I want you to find out what kinds of help are available from OSHA and what kinds of programs OSHA has that will get us certified as a safe company." What can you tell the CEO?

=========== ENDNOTES ===========

1. U.S. Department of Labor. *All About OSHA,* OSHA 2056, 1996 (Revised), p. 2.
2. Ibid., p. 5.
3. Ibid., p. 12.
4. Ibid., pp. 12–13.
5. Ibid., pp. 24–25.
6. Ibid., p. 27.
7. Ibid., p. 28.
8. Ibid., p. 29.
9. Ibid., p. 32.
10. Ibid., pp. 34–35.
11. Ibid., pp. 35–36.
12. Ibid., p. 37.
13. Ibid., pp. 39–40.
14. Ibid., p. 37.

Workers' Compensation

OVERVIEW OF WORKERS' COMPENSATION

The concept of workers' compensation developed as a way to allow injured employees to be compensated appropriately without having to take their employer to court. The underlying rationale for workers' compensation had two aspects: (1) fairness to injured employees, especially those without the resources to undertake legal actions that are often long, drawn out, and expensive; and (2) reduction of costs to employers associated with workplace injuries (e.g., legal, image, and morale costs). Workers' compensation is intended to be a no-fault approach to resolving workplace accidents by rehabilitating injured employees and minimizing their personal losses because of their reduced ability to perform and compete in the labor market.[1] Since its inception as a concept, workers' compensation has evolved into a system that pays out approximately $70 million in benefits and medical costs annually.

Workers' compensation represents a compromise between the needs of employees and the needs of employers. Employees give up their right to seek unlimited compensation for pain and suffering through legal action. Employers award the prescribed compensation (typically through insurance premiums) regardless of the employee's negligence. The theory is that in the long run both employees and employers will benefit more than either would through legal action. Although workers' compensation has reduced the amount of legal action arising out of workplace accidents, it has not completely eliminated legal actions.

Objectives of Workers' Compensation

Workers' compensation laws are not uniform from state to state. In fact, there are significant variations. However, regardless of the language contained in the enabling legislation in a specific state, workers' compensation as a concept has several widely accepted objectives:

1. Replacement of income
2. Rehabilitation of the injured employee
3. Prevention of accidents
4. Cost allocation[2]

The basic premises underlying these objectives are described in the following paragraphs.

Replacement of Income

Employees injured on the job lose income if they are unable to work. For this reason, workers' compensation is intended to replace the lost income adequately and promptly. Adequate income replacement is viewed as replacement of current and future income (minus taxes) at a ratio of two-thirds (in most states). Workers' compensation benefits are required to continue even if the employer goes out of business.

Rehabilitation of the Injured Employee

A basic premise of workers' compensation is that the injured worker will return to work in every case possible, although not necessarily in the same job or career field. For this reason, a major objective of workers' compensation is to rehabilitate the injured employee. The rehabilitation program is to provide the needed medical care at no cost to the injured employee until he or she is pronounced fit to return to work. The program also provides vocational training or retraining as needed. Both components seek to motivate the employee to return to the labor force as soon as possible.

Accident Prevention

Preventing future accidents is a major objective of workers' compensation. The theory underlying this objective is that employers will invest in accident prevention programs in order to hold down compensation costs. Theoretically, the payoff to employers comes in the form of lower insurance premiums that result from fewer accidents.

Cost Allocation

The potential risks associated with different occupations vary. For example, working as a miner is generally considered more hazardous than working as an architect. The underlying principle of cost allocation is to spread the cost of workers' compensation appropriately and proportionately among industries ranging from the most to the least hazardous. The costs of accidents should be allocated in accordance with the accident history of the industry so that high-risk industries pay higher workers' compensation insurance premiums than do low-risk industries.[3]

Who Is Covered by Workers' Compensation?

Workers' compensation laws are written at the state level, and there are many variations among these laws. As a result, it is difficult to make generalizations. Complicating the issue further is the fact that workers' compensation laws are constantly being amended, revised, and rewritten. Additionally, some states make participation in a workers' compensation program voluntary; others excuse employers with fewer than a specified number of employees.

Figure 3–1
Checklist showing who is covered by workers' compensation.
Source: Hammer, W. *Occupational Safety Management and Engineering,* 4th ed. (Upper Saddle River, NJ: Prentice Hall, 1989), p. 36.

Checklist of Who Is Covered by Workers' Compensation

✓ Agricultural employees

✓ Domestic employees

✓ Casual employees

✓ Hazardous work employees

✓ Charitable or religious employees

✓ Employees of small organizations

✓ Railroad and maritime employees

✓ Contractors and subcontractors

✓ Minors

✓ Extraterritoriality employees

In spite of the differences among workers' compensation laws in the various states, approximately 80 percent of the employees in the United States are covered by workers' compensation. Those employees who are not covered or whose coverage is limited vary as the laws vary. However, they can be categorized in general terms shown in the checklist in Figure 3–1.[4]

WORKERS' COMPENSATION INSURANCE

The costs associated with workers' compensation must be borne by employers as part of their overhead. In addition, employers must also ensure that the costs will be paid even if they go out of business. The answer for most employers is workers' compensation insurance.

In most states, workers' compensation insurance is compulsory. Exceptions to this are New Jersey, South Carolina, Texas, and Wyoming. New Jersey allows ten or more employers to form a group and self-insure. Texas requires workers' compensation only for *carriers,* as defined in Title 25, Article 911–A, Section II, Texas state statutes. Wyoming requires workers' compensation only for employers involved in specifically identified *extrahazardous occupations.*

A common thread woven through all of the various compensation laws is the requirement that employers carry workers' compensation insurance. There are three types: state funds, private insurance, and self-insurance. Regardless of the method of coverage chosen, rates can vary greatly from company to company and state to state. Rates are affected by a number of different factors as shown in Figure 3–2.

Figure 3–2
Checklist of factors that affect rates.

Checklist of Factors That Affect Rates

✓ Number of employees

✓ Types of work performed (risk involved)

✓ Accident experience of the employer

✓ Potential future losses

✓ Overhead and profits of the employer

✓ Quality of the employer's safety program

✓ Estimates by actuaries

RESOLUTION OF WORKERS' COMPENSATION DISPUTES

One of the fundamental objectives of workers' compensation is to avoid costly, time-consuming litigation. Whether this objective is being accomplished is questionable. When an injured employee and the employer's insurance company disagree on some aspect of the compensation owed (e.g., weekly pay, length of benefits, degree of disability), the disagreement must be resolved. Most states have an arbitration board for this purpose. Neither the insurance company nor the injured employee is required to hire an attorney. However, many employees do. There are a number of reasons for this. Some don't feel they can adequately represent themselves. Others are fearful of the "big business running over the little guy" syndrome. In any case, workers' compensation litigation is still very common and expensive.

INJURIES AND WORKERS' COMPENSATION

The original workers' compensation concept envisioned compensation for workers who were injured in on-the-job accidents. What constituted an accident varied from state to state. However, all original definitions had in common the characteristics of being *sudden* and *unexpected*. Over the years, the definition of an accident has undergone continual change. The major change has been a trend toward the elimination of the *sudden* characteristic. In many states the gradual onset of an injury or disease as a result of prolonged exposure to harmful substances or a harmful environment can now be considered an accident.

A harmful environment does not have to be limited to its physical components. Psychological factors (such as stress) can also be considered. In fact, the highest rate of growth in workers' compensation claims over the past two decades has been in the area of stress-related injuries. For example, ten years ago, there were 1,282 stress-related claims filed by employees in California. Today this number has increased to over 7,000 claims. This increase in California is indicative of a nationwide trend.

AOE and COE Injuries

Workers' compensation benefits are owed only when the injury arises out of employment (AOE) or occurs in the course of employment (COE). When employees are injured undertaking work prescribed in their job description, work assigned by a supervisor, or work normally expected of employees, they fall into the AOE category. Sometimes, however, different circumstances determine whether the same type of accident is considered to be AOE. For example, say, a soldering technician burns her hand while repairing a printed circuit board that had been rejected by a quality control inspector. This injury would be classified as AOE. Now suppose the same technician brings a damaged printed circuit board from her home stereo to work and burns her hand while trying to repair it. This injury would not be covered because the accident did not arise from her employment. Determining whether an injury should be classified as AOE or COE is often a point of contention in workers' compensation litigation.

Who Is an Employee?

Another point of contention in workers' compensation cases is the definition of the term *employee*. This is an important definition because it is used to determine AOE and COE. A person who is on the company's payroll, receives benefits, and has a supervisor is clearly an employee. However, a person who accepts a service contract to perform a specific task or set of tasks and is not directly supervised by the company is not considered an employee. Although definitions vary from state to state, there are common characteristics. In all definitions, the workers must receive some form of remuneration for work done, and the employer must benefit from this work. Also, the employer must supervise and direct the work, both process and result. These factors—supervision and direction—are what set independent contractors apart from employees and exclude them from coverage. Employers who use independent contractors sometimes require the contractors to show proof of having their own workers' compensation insurance.

DISABILITIES AND WORKERS' COMPENSATION

Injuries that are compensable typically fall into one of four categories: (1) temporary partial disability, (2) temporary total disability, (3) permanent partial disability, and (4) permanent total disability (Figure 3–3). Determining the extent of disability is often a contentious issue. In fact, it accounts for more workers' compensation litigation than any other issue. Further, when a disability question is litigated, the case tends to be complicated because the evidence is typically subjective, and it requires hearing officers, judges, or juries to determine the future.

Temporary Disability

Temporary disability is the state that exists when it is probable that an injured worker, who is currently unable to work, will be able to resume gainful employment with no or only partial disability. Temporary disability assumes that the employee's condition will substantially improve.

There is an important point to remember when considering a temporary disability case. The ability to return to work relates only to work with the company that employed the worker at the time of the accident.

Temporary disability can be classified as either *temporary total disability* or *temporary partial disability*. A temporary total disability classification means the injured worker is incapable of any work for a period of time but is expected to recover fully. Most workers' compensation cases fall in this classification. A temporary partial disability means the injured worker is capable of light or part-time duties. Depending on the extent of the injury, temporary partial disabilities sometimes go unreported. This practice is allowable in some states. It helps employers hold down the cost of their workers' compensation premium. This is similar to not reporting a minor fender bender to your automobile insurance agent.

Permanent Partial Disability

Permanent partial disability is the condition that exists when an injured employee is not expected to recover fully. In such cases, the employee will be able to work again but not at full capacity. Often employees who are partially disabled must be retrained for another occupation.

Figure 3–3
Types of disabilities.

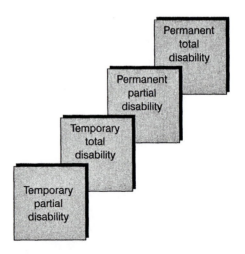

Permanent partial disabilities can be classified as *schedule* or *nonschedule* disabilities. Schedule disabilities are typically the result of nonambiguous injuries such as the loss of a critical but duplicated body part (e.g., arm, ear, hand, finger, or toe). Since such injuries are relatively straightforward, the amount of compensation that they generate and the period of time that it will be paid can be set forth in a standard schedule. Workers' compensation legislation changes continually. Consequently, actual rates are subject to change.

Nonschedule injuries are less straightforward and must be dealt with on a case-by-case basis. Disabilities in this category tend to be the result of head injuries, the effects of which can be more difficult to determine. The amount of compensation awarded and the period over which it is awarded must be determined by studying the evidence. Awards are typically made based on a determination of percent disability. For example, if it is determined that an employee has a 25 percent disability, the employee might be entitled to 25 percent of the income he or she could have earned before the injury with normal career progression factored in.

Four approaches to handling permanent partial disability cases have evolved. Three are based on specific theories, and the fourth is based on a combination of two or more of these theories. The three theories are (1) whole-person theory, (2) wage-loss theory, and (3) loss of wage-earning capacity theory.

Whole-Person Theory

The whole-person theory is the simplest and most straightforward of the theories for dealing with permanent partial disability cases. Once it has been determined that an injured worker's capabilities have been permanently impaired to some extent, this theory is applied like a subtraction problem. What the worker can do after recuperating from the injury is determined and subtracted from what he or she could do before the accident. Factors such as age, education, and occupation are not considered.

Wage-Loss Theory

The wage-loss theory requires a determination of how much the employee could have earned had the injury not occurred. The wages actually being earned are subtracted from what could have been earned, and the employee is awarded a percentage of the difference. No consideration is given to the extent or degree of disability. The only consideration is loss of actual wages.

Loss of Wage-Earning Capacity Theory

The most complex of the theories for handling permanent partial disability cases is the loss of wage-earning capacity theory, because it is based not just on what the employee earned at the time of the accident, but also on what he or she might have earned in the future. Making such a determination is obviously a subjective undertaking. Factors considered include past job performance, education, age, gender, advancement potential at the time of the accident, among others. Once future earning capacity has been determined, the extent to which it has been impaired is estimated, and the employee is awarded a percentage of the difference.

The use of schedules has reduced the amount of litigation and controversy surrounding permanent partial disability cases. This is the good news aspect of schedules. The bad news aspect is that they may be inherently unfair. For example, a surgeon who loses his hand would receive the same compensation as a laborer with the same injury if the loss of a hand is scheduled.

Permanent Total Disability

A permanent total disability exists when an injured employee's disability is such that he or she cannot compete in the job market. This does not necessarily mean that the employee is helpless. Rather, it means an inability to compete reasonably. Handling permanent total disability cases is similar to handling permanent partial disability cases except that certain injuries simplify the process. In most states, permanent total disability can be assumed if certain specified injuries have been sustained (i.e., loss of both eyes or both arms). In some states, compensation is awarded for life. In others, a time period is specified.

MONETARY BENEFITS OF WORKERS' COMPENSATION

The monetary benefits accruing from workers' compensation vary markedly from state to state. The actual amounts are of less importance than the differences among them. Of course, the amounts set forth in schedules change frequently. However, for the purpose of comparison, consider that at one time the loss of a hand in Pennsylvania resulted in an award of $116,245. The same injury in Colorado brought only $8,736.

When trying to determine a scheduled award for a specific injury, it is best to locate the latest schedule for the state in question. One way to do this is to contact the following agency:

U.S. Department of Labor
Employment Standards Administration
Office of State Liaison and Legislative Analysis
Division of State Workers' Compensation Programs
200 Constitution Avenue, N.W.
Washington, D.C. 20210
Internet address: http://www.dol.gov

Death and Burial Benefits

Workers' compensation benefits accrue to the families and dependents of workers who are fatally injured. Typically, the remaining spouse receives benefits for life or until remarriage. However, in some cases, a time period is specified. Dependents typically receive benefits until they reach the legal age of maturity unless they have a condition or circumstances that make them unable to support themselves even after attaining that age.

MEDICAL TREATMENT AND REHABILITATION

All workers' compensation laws provide for payment of the medical costs associated with injuries. Most states provide full coverage, but some limit the amount and duration of coverage. In either case, the employer's maximum financial liability is limited.

The laws also specify who is allowed or required to select a physician for the injured employee. The options can be summarized as follows:

- *Employee selects physician of choice.*   This option is available in Alaska, Arizona, Delaware, Hawaii, Illinois, Kentucky, Louisiana, Maine, Massachusetts, Mississippi, Nebraska, New Hampshire, North Dakota, Ohio, Oklahoma, Oregon, Rhode Island, Texas, the Virgin Islands, Washington, West Virginia, Wisconsin, and Wyoming.
- *Employee selects physician from a list provided by the state agency.*   This option applies in Connecticut, Nevada, New York, and the District of Columbia.
- *Employee selects physician from a list provided by the employer.*   This option applies in Georgia, Tennessee, and Virginia.
- *Employer selects the physician.*   This option applies in Alabama, Florida, Idaho, Indiana, Iowa, Maryland, Montana, New Jersey, New Mexico, North Carolina, South Carolina, and South Dakota.

- *Employer selects the physician, but the selection may be changed by the state agency.*   This option applies in Arkansas, Colorado, Kansas, Minnesota, Missouri, Utah, and Vermont.
- *Employer selects the physician, but after a specified period of time, the employee may choose another.*   This option applies only in Puerto Rico.

Rehabilitation and Workers' Compensation

Occasionally an injured worker will need rehabilitation before he or she can return to work. There are two types of rehabilitation: medical and vocational. Both are available to workers whose ability to make a living is inhibited by physical and/or mental work-related problems.

Medical rehabilitation consists of providing whatever treatment is required to restore to the extent possible any lost ability to function normally. This might include such services as physical therapy or the provision of prosthetic devices. Vocational rehabilitation involves providing the education and training needed to prepare the worker for a new occupation. Whether the rehabilitation services are medical or vocational in nature or both, the goal is to restore the injured worker's capabilities to the level that existed before the accident.

ADMINISTRATION AND CASE MANAGEMENT

It is not uncommon for minor injuries to go unreported. The employee might be given the rest of the day off or treated with first aid and returned to work. This is done to avoid time-consuming paperwork and to hold down the cost of workers' compensation insurance. However, if an accident results in a serious injury, several agencies must be notified. What constitutes a serious injury, like many workers' compensation issues, can differ from state to state. However, as a rule, an injury is serious if it requires more than 24 hours of active medical treatment (this does not include passive treatment such as observation). Of course, a fatality, a major disfigurement, or the loss of a limb or digit is also considered serious and must be reported.

At a minimum, the company's insurer, the state agency, and the state's federal counterpart must be notified. Individual states may require that additional agencies be notified. All establish a time frame within which notification must be made. Once the notice of injury has been filed, there is typically a short period before the victim or dependents can begin to receive compensation unless in-patient hospital care is required. However, when payments do begin to flow, they are typically retroactive to the date of the injury.

State statutes also provide a maximum time period that can elapse before a compensation claim is filed. The notice of injury does not satisfy the requirement of filing a claim notice. The two are separate processes. The statute of limitations on claim notices varies from state to state. However, most limit the period to no more than a year except in cases of work-related diseases in which the exact date of onset cannot be determined.

All such activities—filing injury notices, filing claim notices, arriving at settlements, and handling disputes—fall under the collective heading of administration and case management. Most states have a designated agency that is responsible for administration and case management. In addition, some states have independent boards that conduct hearings and/or hear appeals when disputes arise.

Once a workers' compensation claim is filed, an appropriate settlement must be reached. Three approaches can be used to settle a claim: (1) direct settlement, (2) agreement settlement, and (3) public hearing. The first two are used in uncontested cases, the third in contested cases.

1. *Direct settlement.* The employer or its insurance company begins making what it thinks are the prescribed payments. The insurer also sets the period over which payments will be made. Both factors are subject to review by the designated state agency.

2. *Agreement settlement.* The injured employee and the employer or its insurance company work out an agreement on how much compensation will be paid and for how long. Such an agreement must be reached before compensation payments begin. Typically, the agreement is reviewed by the designated state administrative agency. In cases where the agency disapproves the agreement, the worker continues to collect compensation at the agreed-on rate until a new agreement can be reached. If this is not possible, the case becomes a contested case.

3. *Public hearing.* If an injured worker feels he or she has been inadequately compensated or unfairly treated, a hearing can be requested. Such cases are known as *contested cases.* The hearing commission reviews the facts surrounding the case and renders a judgment concerning the amount and duration of compensation. Should the employee disagree with the decision rendered, civil action through the courts is an option.

PROBLEMS WITH WORKERS' COMPENSATION

There are serious problems with workers' compensation in the United States. On the one hand, there is evidence of abuse of the system. On the other hand, many injured workers who are legitimately collecting benefits suffer a substantial loss of income. Complaints

---

Checklist of "Red Flag" Factors

✔ The person filing the claim is never home or available by telephone or has an unlisted telephone number.

✔ The injury in question coincides with a layoff, termination, or plant closing.

✔ The person filing the claim is active in sports.

✔ The person filing the claim has another job.

✔ The person filing the claim is in line for early retirement.

✔ The rehabilitation report contains evidence that the person filing the claim is maintaining an active lifestyle.

✔ No organic basis exists for disability. The person filing the claim appears to have made a full recovery.

✔ The person filing the claim receives all mail at a post office box and will not divulge a home address.

✔ The person filing the claim is known to have skills such as carpentry, plumbing, or electrical that could be used to work on a cash basis while feigning a disability.

✔ There are no witnesses to the accident in question.

✔ The person filing the claim has relocated out of the state or out of the country.

✔ Demands for compensation are excessive.

✔ The person filing the claim has a history of filing.

✔ Doctor's reports are contradictory in nature.

✔ A soft-tissue injury is claimed to have produced a long-term disability.

✔ The injury in question occurred during hunting season.

Figure 3–4
Checklist of "red flag" factors.
Source: Kizorek, B. "Video Surveillance and Workers' Comp Fraud," *Occupational Hazards,* February 1994, p. 28.

about workers' compensation are common from all parties involved with the system (employers, employees, and insurance companies). A major issue is psychological claims (primarily, stress). Such claims have increased rapidly, while physiological claims have leveled off somewhat.

In 1980, there were so few stress claims they were not even recorded as a separate category. Today, they represent a major and costly category.

Stress claims are more burdensome than physical claims because they are typically reviewed in an adversarial environment. This leads to the involvement of expert medical witnesses and attorneys. As a result, even though the benefits awarded for stress-related injuries are typically less than those awarded for physical injuries, the cost of stress claims is often higher because of the litigation.

The most fundamental problem with workers' compensation is that it is not fulfilling its objectives. Lost income is not being adequately replaced, the number of accidents has not decreased, and the effectiveness of cost allocation is questionable. Clearly, the final chapter on workers' compensation has not yet been written.

SPOTTING WORKERS' COMPENSATION FRAUD/ABUSE

There is evidence of waste, fraud, and abuse of the system in all states that have passed workers' compensation laws. However, the public outcry against fraudulent claims is making states much less tolerant of, and much more attentive to, abuse. For example, the Ohio legislature passed a statute that allows criminal charges to be brought against employees, physicians, and lawyers who give false information in a workers' compensation case. This is a positive trend. However, even these measures will not completely eliminate abuse.

For this reason, it is important to know how to spot employees who are trying to abuse the system by filing fraudulent workers' compensation claims. Following are some factors that should cause employers to view claims with suspicion. However, just because one or more of these factors is present does not automatically mean that an employee is attempting to abuse the system. Rather, the factors summarized in Figure 3–4 are simply factors that should raise cautionary flags.[5]

These factors can help organizations spot employees who may be trying to abuse the workers' compensation system. It is important to do so because legitimate users of the system are hurt, as are their employers, by abusers of the system. If one or more of these factors is present, employers should investigate the claim carefully before proceeding.

COST REDUCTION STRATEGIES

All managers, engineers, and technologists are responsible for helping their organizations hold down workers' compensation costs. Of course, the best way to accomplish this goal is to maintain a safe and healthy workplace, thereby preventing the injuries that drive the costs up. This section presents numerous other strategies that have proven effective in reducing workers' compensation costs after injuries have occurred, which happen in even the safest environments.

General Strategies

Regardless of the type of organization, there are several rules of thumb that can help reduce workers' compensation claims. These general strategies are as follows (see Figure 3–5):

Specific Strategies

In addition to the general strategies just presented, there are numerous specific cost-containment strategies that have proven to be effective as shown in Figure 3–5.

Figure 3–5
Checklist of cost reduction strategies.

Checklist of Cost Reduction Strategies

General Strategies

✓ Stay in touch with injured employees.

✓ Have a return-to-work program and use it.

✓ Determine the cause of the accident and eliminate it.

Specific Strategies

✓ Cultivate job satisfaction.

✓ Make safety part of the organizational culture.

✓ Have a systematic cost reduction program.

✓ Use integrated managed care.

APPLICATION SCENARIOS

1. You have been chosen to give a guest lecture to a college class on the subject of workers' compensation. The professor wants you to speak about the following issues: (a) rationale for workers' compensation, (b) objectives of workers' compensation, and (c) whether workers' compensation really works. What will you tell the students?

2. An employee on the shop floor has been injured and treated with on-site first aid. She wants to go back to work, but her supervisor and the personnel manager are in disagreement over whether the injury should be considered serious. You have been asked to mediate the argument. How will you distinguish a minor injury from a serious injury?

3. You have been asked to chair an ad hoc committee to identify factors in your organization that are contributing to high workers' compensation costs. Develop a checklist of factors to consider.

4. You have been asked to serve on a workers' compensation arbitration panel. The panel's first task is to decide if the individual in the following case qualifies as an employee:

   Jane Andrews used to work fulltime for McKnight Publishing as a copy editor. With the advent of e-mail and the Internet, Andrews is able to work at home and communicate electronically. She now has a contract with McKnight and works out of her home. Her next-door neighbor is her former supervisor at McKnight, Jill Mandis. Andrews still sends her work to Mandis. In fact, Mandis often visits Andrews in the evenings to review work in progress. Recently, Andrews had an automobile accident while driving to McKnight for a face-to-face meeting with an author. When the accident occurred, Andrews was talking to a secretary at McKnight on her car phone. Should Andrews be considered a McKnight employee? Is McKnight responsible for Andrews' medical costs and the damage to her automobile?

5. You are convinced that your company is the victim of workers' compensation fraud and abuse. Develop a checklist of factors that all professionals in the company can use to help them spot fraud and abuse.

ENDNOTES

1. Society of Manufacturing Engineers. *Manufacturing Management* (Dearborn, MI: Society of Manufacturing Engineers, 1988), pp. 12–24.
2. Ibid.
3. Ibid.
4. Hammer, W. *Occupational Safety Management and Engineering,* 4th ed. (Upper Saddle River, NJ: Prentice Hall, 1989), p. 36.
5. Kizorek, B. "Video Surveillance and Workers' Comp Fraud," *Occupational Hazards,* February 1994, p. 28.

CHAPTER FOUR

Ergonomics Hazards and Repetitive Strain Injuries

The history of workplace development in the Western world is characterized by jobs and technologies designed to improve processes and productivity. All too often in the past, little or no concern was given to the impact of the process or technology on workers. As a result, processes and machines have sometimes been unnecessarily dangerous. Another result has been that new technologies have sometimes failed to live up to expectations. This is because, even in the age of global competition, human involvement in work processes is still the key to the most significant and enduring productivity improvements. If a machine or system is uncomfortable, difficult, overly complicated, or dangerous, human workers will not be able to derive the full benefit of its use.

The proliferation of uncomfortable and dangerous workplace conditions, whether created by job design or unfriendly technologies, is now widely recognized as harmful to productivity, quality, and worker safety and health. The advent of the science of ergonomics is making the workplace more physically friendly. This, in turn, is making the workplace a safer and healthier place.

ERGONOMICS DEFINED

Minimizing the amount of physical stress in the workplace requires continuous study of the ways in which people and technology interact. The insight learned from this study must then be used to improve the interaction. This is a description of the science of ergonomics. For the purpose of this book, ergonomics is defined as follows:

> *Ergonomics is the science of conforming the workplace and all of its elements to the worker.*

The word *ergonomics* is derived from the Greek language. *Ergon* is Greek for *work;* *nomos* means *laws.* Therefore, in a literal sense, ergonomics means work laws. In practice, it consists of the scientific principles (laws) applied in minimizing the physical stress associated with the workplace (work). Figure 4–1 summarizes some of the widely accepted benefits of ergonomics.

FACTORS ASSOCIATED WITH PHYSICAL STRESS[1]

In a book translated from work originally produced by the Swedish Work Environment Fund, the National Safety Council lists eight variables that can influence the amount of physical stress experienced on the job:

1.	Sitting	Standing
2.	Stationary	Moveable/mobile
3.	Large demand for strength/power	Small demand for strength/power
4.	Good horizontal work area	Bad horizontal work area
5.	Good vertical work area	Bad vertical work area
6.	Nonrepetitive motion	Repetitive motion
7.	Low surface	High surface
8.	No negative environmental factors	Negative environmental factors

The following paragraphs summarize how these factors compare in terms of physical stress.

Sitting vs. Standing

Generally speaking, sitting is less stressful than standing. Standing for extended periods, particularly in one place, can produce unsafe levels of stress on the back, legs, and feet. Although less so than standing, sitting can be stressful unless the appropriate precautions are taken. These precautions include proper posture, a supportive back rest, and frequent standing/stretching movement.

Stationary vs. Mobile

Stationary jobs are those done primarily at one workstation. Of course, even these jobs involve movement at the primary workstation and occasional movement to other areas.

Figure 4–1
Checklist of the benefits of ergonomics.

Checklist of the Benefits of Ergonomics

✓ Improved health and safety for workers

✓ Higher morale throughout the workplace

✓ Improved quality

✓ Improved productivity

✓ Improved competitiveness

✓ Decreased absenteeism and turnover

✓ Fewer workplace injuries/health problems

Mobile jobs, on the other hand, require continual movement from one station to another. The potential for physical stress increases with stationary jobs when workers fail to take such precautions as periodically standing/stretching/moving. The potential for physical stress increases with mobile jobs when workers carry materials as they move from station to station.

Large vs. Small Demand for Strength/Power

In classifying jobs by these two criteria, it is important to understand that repeatedly moving small amounts of weight over a period of time can have a cumulative effect equal to the amount of stress generated by moving a few heavy weights. Regardless of whether the stress results from lifting a few heavy objects or repeated lifting of lighter objects, jobs that demand larger amounts of strength/power are generally more stressful than those requiring less.

Good vs. Bad Horizontal Work Area

A good horizontal work area is one that is designed and positioned so that it does not require the worker to bend forward or to twist the body from side to side. Horizontal work areas that do require these movements are bad. Bad horizontal work surfaces increase the likelihood of physical stress.

Good vs. Bad Vertical Work Area

Good vertical work areas are designed and positioned so that workers are not required to lift their hands above their shoulders or bend down in order to perform any task. Vertical work areas that do require these movements are bad. Bad vertical work areas increase the likelihood of physical stress.

Nonrepetitive vs. Repetitive Motion

Jobs that are repetitive involve short-cycle motion that is repeated continually. Nonrepetitive jobs involve a variety of tasks that are not, or only infrequently, repeated. Repetition can lead to monotony and boredom. When this happens, the potential for physical stress increases.

Low vs. High Surface Contact

Surface stress can result from contact with hard surfaces such as tools, machines, and equipment. High surface contact jobs tend to be more stressful in a physical sense than are low surface contact jobs.

Presence vs. Absence of Environmental Factors

Generally, the more environmental factors with which a worker has to contend on the job, the more stressful the job. For example, personal protective equipment, although conducive to reducing environmental hazards, can increase the amount of physical stress associated with the job.

OSHA'S ERGONOMICS GUIDELINES

OSHA publishes guidelines for general safety and health program management.[2] OSHA's ergonomic guidelines are voluntary and are designed to provide employers with the information and guidance needed to meet their obligations under the OSHAct regarding ergonomics.

Checklist of the Worksite Analysis for Ergonomic Hazards

_✓_ Are tasks being performed that involve unnatural movements?

_✓_ Are tasks being performed that involve frequent manual lifting?

_✓_ Are tasks being performed that involve excessive wasted motion?

_✓_ Are tasks being performed that involve unnatural or uncomfortable postures?

Figure 4–2
Checklist of worksite analysis for ergonomic hazards.

OSHA also publishes guidelines designed specifically for the meat-packing industry.[3] These specific guidelines represent a model for guidelines that are likely to be developed for other specific industries. Meat packing was singled out for attention because of the high incidence of cumulative trauma disorders (CTDs) associated with meat packing. CTDs are injuries that result from an accumulation of repetitive motion stress. For example, using scissors continually over time can cause a CTD in the hand and wrist.

WORKSITE ANALYSIS PROGRAM FOR ERGONOMICS[*]

Although complex analyses are best performed by a professional ergonomist, the "ergonomic team"—or any qualified person—can use this program to conduct a worksite analysis and identify stressors in the workplace. The purpose of the following information is to give a starting point for finding and eliminating those tools, techniques, and conditions that may be the source of ergonomic problems.

In addition to analyzing current workplace conditions, planned changes to existing and new facilities, processes, materials, and equipment should be analyzed to ensure that changes made to enhance production will also reduce or eliminate ergonomic risk factors. As emphasized before, this program should be adapted to each individual workplace.

Figure 4–2 is a checklist that can be used for analyzing the workplace for ergonomic hazards. Figure 4–3 is a checklist for CTD hazards, and Figure 4–4 is a checklist for back-related hazards.

HAZARD PREVENTION AND CONTROL

Engineering solutions, where feasible, are the preferred method for ergonomic hazard prevention and control. The focus of an ergonomics program is to make the job fit the person—not to make the person fit the job. This is accomplished by redesigning the workstation, work methods, or tool to reduce the demands of the job, including high force, repetitive motion, and awkward postures. Following are some examples of engineering controls that have proven to be effective and achievable.

Workstation Design

Workstations should be designed to accommodate the persons who actually use them; it is not sufficient to design for the average or typical worker. Workstations should be easily adjustable and should be either designed or selected to fit a specific task, so that they are comfortable for the workers who use them. The work space should be large enough to

---

[*]Adapted from OSHA 3123, 1991 (reprinted).

Figure 4–3
Checklist of CTD risk factors for
the upper extremities.

> **Checklist of CTD Risk Factors**
>
> ✓ Repetitive and/or prolonged activities
>
> ✓ Forceful exertions, usually with the hands (including pinch grips)
>
> ✓ Prolonged static postures
>
> ✓ Awkward postures of the upper body, including reaching above the shoulders or behind the back, and twisting the wrists and other joints to perform tasks
>
> ✓ Continual physical contact with work surfaces (for example, contact with edges)
>
> ✓ Excessive vibration from power tools
>
> ✓ Cold temperatures
>
> ✓ Inappropriate or inadequate hand tools

allow for the full range of required movements, especially where knives, saws, hooks, and similar tools are used.

Design of Work Methods

Traditional work method analysis considers static postures and repetition rates. This should be supplemented by addressing the force levels and the hand and arm postures involved. The tasks should be altered to reduce these and the other stresses associated with CTDs. The results of such analyses should be shared with the health-care providers to assist in compiling lists of light-duty and high-risk jobs.

Tool Design and Handles

Tools should be selected and designed to minimize the risks of upper extremity CTDs and back injuries. In any tool design, a variety of sizes should be available. Examples of criteria for selecting tools include the following:

■ Designing tools to be used by either hand, or providing tools for both left- and right-handed workers.

■ Using tools with triggers that depress easily and are activated by two or more fingers.

Figure 4–4
Checklist of back disorder risk
factors.

> **Checklist of Back Disorder Risk Factors**
>
> ✓ Bad body mechanics such as continued bending over at the waist, continued lifting from below the knees or above the shoulders, and twisting at the waist, especially while lifting
>
> ✓ Lifting or moving objects of excessive weight or asymmetric size
>
> ✓ Prolonged sitting, especially with poor posture
>
> ✓ Lack of adjustable chairs, footrests, body supports, and work surfaces at workstations
>
> ✓ Poor grips on handles
>
> ✓ Slippery footing

■ Using handles and grips that distribute the pressure over the fleshy part of the palm, so that the tool does not dig into the palm.

■ Designing/selecting tools for minimum weight; counterbalancing tools heavier than one or two pounds.

■ Selecting pneumatic and power tools that exhibit minimal vibration and maintaining them in accordance with manufacturer's specifications or with an adequate vibration-monitoring program. Wrapping handles and grips with insulation material (other than wraps provided by the manufacturer for this purpose) is normally *not* recommended, as it may interfere with a proper grip and increase stress.

MEDICAL MANAGEMENT PROGRAM*

An effective medical management program for cumulative trauma disorders is essential to the success of an employer's ergonomic program in industries with a high incidence of CTDs. It is not the purpose of these guidelines to dictate medical practice for an employer's health-care providers. Rather, they describe the elements of a medical management program for CTDs to ensure early identification, evaluation, and treatment of signs and symptoms; to prevent their recurrence; and to aid in their prevention. Medical management of CTDs is a developing field, and health-care providers should monitor developments on the subject. These guidelines represent the best information currently available.

In an effective ergonomics program, health-care providers should be part of the ergonomics team interacting and exchanging information routinely to prevent and treat CTDs properly. The major components of a medical management program for the prevention and treatment of CTDs are trained first-level health-care providers, health surveillance, employee training and education, early reporting of symptoms, appropriate medical care, accurate record keeping, and quantitative evaluation of CTD trends throughout the plant.

TRAINING AND EDUCATION**

An essential element for an effective ergonomics program is training and education. The purpose of training and education is to ensure that employees are sufficiently informed about the ergonomic hazards to which they may be exposed and thus able to participate actively in their own protection.

Training and education allow managers, supervisors, and employees to understand the hazards associated with a job or process, their prevention and control, and their medical consequences. A training program should include all affected employees, engineers and maintenance personnel, supervisors, and health-care providers.

The program should be designed and implemented by qualified persons. Appropriate special training should be provided for personnel responsible for administering the program. The program should be presented in language and at a level of understanding appropriate for the individuals being trained. It should provide an overview of the potential risk of illnesses and injuries, their causes and early symptoms, the means of prevention, and treatment.

The program should also include a means for adequately evaluating its effectiveness. This might be achieved by using employee interviews, testing, and observing work

---

*Adapted from OSHA 3123, 1991 (reprinted).
**Adapted from OSHA 3123, 1991 (reprinted).

practices, to determine if those who received the training understand the material and the work practices to be followed.

COMMON INDICATORS OF PROBLEMS

Does my company have ergonomic problems? Are injuries/illnesses occurring because too little attention is paid to ergonomic factors? These are questions that modern safety and health professionals should ask themselves. But how does one answer such questions? According to the National Safety Council, the factors discussed in the following paragraphs can be examined to determine if ergonomic problems exist in a given company.[4]

Apparent Trends in Accidents and Injuries

By examining accident reports, record-keeping documents such as OSHA Form 200 (Log and Summary of Occupational Injuries and Illnesses), first aid logs, insurance forms, and other available records of illnesses or injuries, safety and health professionals can identify trends if they exist. A pattern or a high incidence rate of a specific type of injury typically indicates that an ergonomic problem exists.

Incidence of Cumulative Trauma Disorders

Factors associated with CTDs include a high level of repetitive work, greater than normal levels of hand force, awkward posture, high levels of vibration, high levels of mechanical stress, extreme temperatures, and repeated hand-grasping/pinch-gripping. By observing the workplace and people at work, safety and health professionals can determine the amount of exposure that employees have to these factors and the potential for ergonomics-related problems.

Absenteeism and High Turnover Rates

High absentee rates and high turnover rates can be indicators of ergonomic problems. People who are uncomfortable on the job to the point of physical stress are more likely to miss work and/or leave for less stressful conditions.

Employee Complaints

A high incidence of employee complaints about physical stress or poor workplace design can indicate the presence of ergonomic problems.

Employee-Generated Changes

Employees tend to adapt the workplace to their needs. The presence of many workplace adaptations, particularly those intended to decrease physical stress, can indicate the presence of ergonomic problems. Have employees added padding, modified personal protective equipment, brought in extra lighting, or made other modifications? Such employee-generated changes may be evidence of ergonomic problems.

Poor Quality

Poor quality, although not necessarily caused by ergonomic problems, can be the result of such problems. Poor quality is at least an indicator that there may be ergonomic problems. Certainly, poor quality is an indicator of a need for closer inspection.

Manual Material Handling

The incidence of musculoskeletal injuries is typically higher in situations that involve a lot of manual material handling. Musculoskeletal injuries increase significantly when the job involves one or more of the following: lifting large objects, lifting bulky objects, lifting objects from the floor, and lifting frequently. When such conditions exist, the company has ergonomic problems.

IDENTIFYING SPECIFIC ERGONOMIC PROBLEMS

Specific ergonomic problems are identified by conducting a task analysis of the job in question. Figure 4–5 lists the types of problems that can be identified by a thorough task analysis.[5] The National Safety Council recommends using one or more of the following approaches for conducting a task analysis:

- *General observation.*   Observing a worker or workers performing the task(s) in question can be an effective task analysis technique. The effectiveness is usually enhanced if the workers are not aware that they are being observed. When observing employees at work, be especially attentive to tasks requiring manual material handling and repetitive movements.

- *Questionnaires and interviews.*   These can be used for identifying ergonomic problems. Questionnaires are easier to distribute, tabulate, and analyze, but interviews generally provide more in-depth information.

- *Videotaping and photography.*   Videotaping technology has simplified the process of task analysis considerably. Videotaping records the work being observed as it is done, it is silent so it is not intrusive, and such capabilities as freeze and playback enhance the observer's analysis capabilities significantly. Photography can also enhance the observer's analysis capabilities by recording each motion or step involved in performing a task. If photography is used, be aware that flashes can be disruptive. High-speed film will allow you to make photographs without using a flash.

- *Drawing or sketching.*   Making a neat sketch of a workstation or a drawing showing workflow can help identify problems. Before using a drawing or sketch as part of a task analysis, make sure that it is accurate. Ask an employee who is familiar with the area or process sketched to check the drawing.

- *Measuring the work environment.*   Measurements can help identify specific ergonomic problems. How far must a worker carry the material manually? How high

Figure 4-5
Problems that can be pinpointed by a task analysis.

> **Checklist of Problems That Can Be Identified by Task Analysis**
>
> ✓ Tasks that involve potentially hazardous movements
> ✓ Tasks that involve frequent manual lifting
> ✓ Tasks that involve excessive wasted motion or energy
> ✓ Tasks that are part of a poor operations flow
> ✓ Tasks that require unnatural or uncomfortable posture
> ✓ Tasks with high potential for psychological stress
> ✓ Tasks with a high fatigue factor
> ✓ Tasks that could/should be automated
> ✓ Tasks that involve or lead to quality control problems

does a worker have to lift an object? How much does an object weigh? How often is a given motion repeated? Answers to these and similar questions can enhance the effectiveness of the analysis process.

■ *Understanding the ergonomics of aging.*   When identifying specific ergonomic problems in the workplace, don't overlook the special challenges presented by aging workers. A good ergonomics program adapts the job to the person. Since nearly 30 percent of the workforce is 45 years of age or older, organizations must be prepared to adapt workstations to employees whose physical needs are different from those of their younger counterparts.

In adapting workstations and processes for employees who are 45 or older, keep the following rules of thumb in mind:[6]

■ Nerve conduction velocity, hand-grip strength, muscle mass, range of motion, and flexibility all begin to diminish about age 45.

■ Weight and mass tend to increase through about the early fifties.

■ Height begins to diminish beginning around age 30.

■ Lower back pain is more common in people 45 years of age and older.

■ Visual acuity at close range diminishes with age.

These rules of thumb mean that safety and health professionals cannot take a "one-size-fits-all" approach to ergonomics. Adaptations for older workers must be individualized and should take aging factors into account.

VDTs AND ERGONOMICS: REDUCING HAZARDS

The video display terminal, primarily because of the all-pervasive integration of personal computers in the workplace, is now the most widely used piece of office equipment. This fact coupled with the ergonomic hazards associated with VDTs have created a whole new range of concerns for safety and health professionals. Using ergonomics to design a workspace will make it easier, safer, more comfortable, and more efficient to use. Following are some strategies that can be used to reduce the hazards associated with VDTs:[7]

■ *Arrange the keyboard properly.*   It should be located in front of the user, not to the side. Body posture and the angle formed by the arms are critical factors (see Figure 4-6).

■ *Adjust the height of the desk.*   Taller employees often have trouble working at *average* height desks. Raising the desk with wooden blocks can solve this problem.

■ *Adjust the tilt of the keyboard.*   The rear portion of the keyboard should be lower than the front.

■ *Encourage employees to use a soft touch on the keyboard and when clicking a mouse.*   A hard touch increases the likelihood of injury.

■ *Encourage employees to avoid wrist resting.*   Resting the wrist on any type of edge can increase pressure on the wrist.

■ *Place the mouse within easy reach.*   Extending the arm to its full reach increases the likelihood of injury.

■ *Remove dust from the mouse ball cavity.*   Dust can collect, making it difficult to move the mouse. Blowing out accumulated dust once a week will keep the mouse easy to manipulate.

■ *Locate the VDT at a proper height and distance.*   The VDT's height should be such that the top line on the screen is slightly below eye level. The optimum distance between the VDT and user will vary from employee to employee, but it will usually be between 16 and 32 inches.

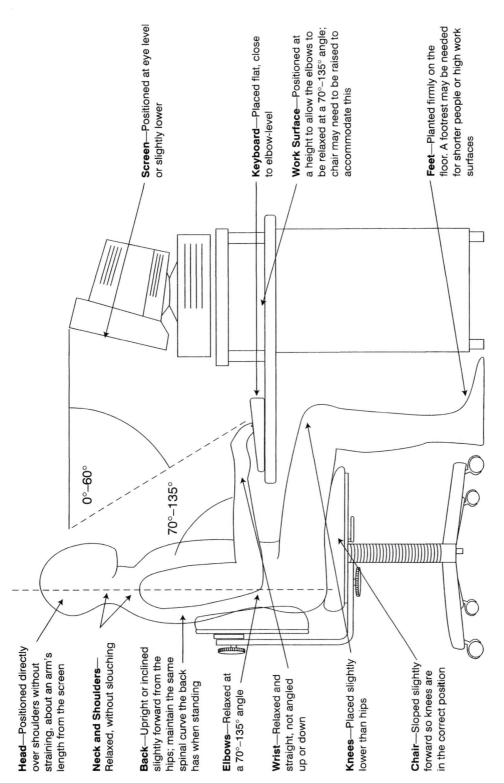

Head—Positioned directly over shoulders without straining, about an arm's length from the screen

Neck and Shoulders—Relaxed, without slouching

Back—Upright or inclined slightly forward from the hips; maintain the same spinal curve the back has when standing

Elbows—Relaxed at a 70°–135° angle

Wrist—Relaxed and straight, not angled up or down

Knees—Placed slightly lower than hips

Chair—Sloped slightly forward so knees are in the correct position

Screen—Positioned at eye level or slightly lower

Keyboard—Placed flat, close to elbow-level

Work Surface—Positioned at a height to allow the elbows to be relaxed at a 70°–135° angle; chair may need to be raised to accommodate this

Feet—Planted firmly on the floor. A footrest may be needed for shorter people or high work surfaces

0°–60°

70°–135°

Figure 4-6
Ergonomics of VDTs. The diagram on the left highlights optimal postures and positions for the computer user.

- *Minimize glare.*   Glare from a VDT can cause employees to adopt harmful postures. Changing the location of the VDT, using a screen hood, and closing or adjusting blinds and shades can minimize glare.
- *Reduce lighting levels.*   Reducing the lighting level in the area immediately around the VDT can eliminate vision strain.
- *Dust the VDT screen.*   VDT screens are magnets to dust. Built-up dust can make the screen difficult to read, contributing to eye strain.
- *Eliminate telephone cradling.*   Cradling a telephone receiver between an uplifted shoulder and the neck while typing can cause a painful disorder called cervical radiculopathy (compression of the cervical vertebrae in the neck). Employees who need to talk on the telephone while typing should wear a headphone.
- *Require typing breaks.*   Continuous typing for extended periods should be avoided. Repetitive strain injuries are cumulative. Breaking up the repetitive motion in question (typing and clicking) can help prevent the accumulation of strain.

REPETITIVE STRAIN INJURY (RSI) DEFINED

The personal computer has become an all-pervasive and universal work tool. Jobs from the shop floor to the executive office now involve frequent, repetitive computer use. This means that people in the workplace are typing and clicking at an unprecedented pace. Frequent and, for some, constant computer use have led to an explosion of injuries heretofore seen mostly in the meat-packing industry. Collectively, these injuries are known as repetitive strain injuries or RSIs.

Definition

RSI is an umbrella term that covers a number of cumulative trauma disorders (CTDs) caused by forceful or awkward hand movements repeated frequently over time. Other aggravating factors include poor posture, an improperly designed workstation, and job stress. RSIs occur to the muscles, nerves, and tendons of the hands, arms, shoulders, and neck.

CLASSIFICATIONS OF RSIs[8]

For years, RSI has been incorrectly referred to as *carpal tunnel syndrome,* which, in reality, is actually one type of RSI. This is like referring to all trees as oaks. Figure 4-7 is a checklist of the most common RSIs organized into four broad classifications.

Muscle and Tendon Disorders

Tendons connect muscles to bones. They can accommodate very little in the way of stretching and are prone to injury if overused. Overworking a tendon can cause small tears in it. These tears can become inflamed and cause intense pain. This condition is known as tendinitis.

Myofacial muscle damage can also be caused by overexertion. It manifests itself in soreness that persists even when resting. Muscles may burn and be sensitive to the touch. When sore muscles become inflamed and swell, the symptoms are aggravated even further by nerve compression. Tendons that curve around bones are encased in protective coverings called *sheathes.* Sheathes contain a lubricating substance known as synovial fluid. When tendons rub against the sheath too frequently, friction is produced. The body responds by producing additional synovial fluid. Excess build-up of this fluid can cause swelling which, in turn, causes pressure on the surrounding nerves, causing a condition known as tenosynovitis.

Figure 4-7
RSI checklist—types of injury by classification.

Checklist of Injury Types by Classification

Muscle and Tendon Disorders

✓ Tendinitis

✓ Muscle damage (myofacial)

✓ Tenosynovitis

✓ Stenosing tenosynovitis

- DeQuervain's disease
- Trigger finger (flexor tenosynovitis)

✓ Shoulder tendinitis

✓ Bicipital tendinitis

✓ Rotator cuff tendinitis

✓ Forearm tendinitis

- Flexor carpi radialis tendinitis
- Extensor tendinitis
- Flexor tendinitis

✓ Epicondylitis

✓ Ganglion cysts

Cervical Radiculopathy

Tunnel Syndromes

✓ Carpal tunnel syndrome

✓ Radial tunnel syndrome

✓ Sulcus ulnaris syndrome

✓ Cubital tunnel syndrome

✓ Guyons canal syndrome

Nerve and Circulation Disorders

✓ Thoracic outlet syndrome

✓ Raynaud's disease

Chronic tenosynovitis is known as *stenosing tenosynovitis* of which there are two types: DeQuervain's disease and flexor tenosynovitis (trigger finger). DeQuervain's disease affects the tendon at the junction of the wrist and thumb. It causes pain when the thumb is moved or when the wrist is twisted. Flexor tenosynovitis involves the locking of a digit in a bent position, hence the term *trigger finger*. However, it can occur in any finger.

Shoulder tendinitis is of two types: bicipital and rotator cuff tendinitis. Bicipital tendinitis occurs at the shoulder joint where the bicep muscle attaches. The rotator cuff is a group of muscles and tendons in the shoulder that move the arm away from the body and turn it in and out. Pitchers in baseball and quarterbacks in football often experience rotator cuff tendinitis.

Forearm tendinitis is of three types: flexor carpi radialis tendinitis, extensor tendinitis, and flexor tendinitis. Flexor carpi radialis tendinitis causes pain in the wrist at the base of the thumb. Extensor tendinitis causes pain in the muscles in the top of the hand, making it difficult to straighten the hands. Flexor tendinitis causes pain in the fingers, making them difficult to bend.

Epicondylitis and *ganglion cysts* are two muscle and tendon disorders. Epicondylitis (lateral) affects the outside of the elbow, whereas epicondylitis (medial) affects the inside. The common term for this disorder is tennis "elbow." Ganglion cysts grow on the tendon, tendon sheath, or synovial lining, typically on top of the hand, on the nail bed, above the wrist, or on the inside of the wrist.

Cervical Radiculopathy

This disorder is most commonly associated with holding a telephone receiver on an upraised shoulder while typing. This widely practiced act can cause compression of the cervical discs in the neck, making it painful to turn the head. Putting the body in an unnatural posture while using the hands is always dangerous.

Tunnel Syndromes

Tunnels are conduits for nerves that are formed by ligaments and other soft tissues. Damage to the soft tissues can cause swelling that compresses the nerves that pass through the tunnel. These nerves are the median, radial, and ulnar nerves that pass through a tunnel in the forearm and wrist. Pain experienced with tunnel injuries can be constant and intense. In addition to pain, people with a tunnel injury might experience numbness, tingling, and a loss of gripping power. The most common tunnel syndromes are carpal tunnel syndrome, radial tunnel syndrome, sulcus ulnaris syndrome, cubital tunnel syndrome, and guyon's canal syndrome.

Nerve and Circulation Disorders

When friction or inflammation cause swelling, both nerves and arteries can be compressed, restricting the flow of blood to muscles. This can cause a disorder known as *thoracic outlet syndrome*. The symptoms of this disorder are pain in the entire arm, numbness, coldness, and weakness in the arm, hand, and fingers.

If the blood vessels in the hands are constricted, *Raynaud's disease* can result. Symptoms include painful sensitivity, tingling, numbness, coldness, and paleness in the fingers. It can affect one or both hands. This disorder is also known as *vibration syndrome* because it is associated with vibrating tools.

PREVENTING RSI[9]

The best way to prevent RSI is to make employees aware of the hazards that can cause it. These hazards include poor posture at the workstation, inappropriate positioning of the hands and arms, a heavy hand on a keyboard or mouse, and any other act that repeatedly puts the body in an unnatural posture while using the hands. Ergonomically sound workstations can help prevent RSI, especially when they can be modified to fit the individual employee. However, even the best ergonomic design cannot prevent a heavy hand on the keyboard or mouse. Consequently, ergonomics is only part of the answer. Following are some prevention strategies that can be applied in any organization:

1. *Teach employees the warning signs.* RSI occurs cumulatively over time. It sneaks up on people. Employees should be aware of the following warning signs: weakness in the hands or forearms, tingling, numbness, heaviness in the hands, clumsiness, stiffness, lack of control over the fingers, cold hands, and tenderness to the touch.

2. *Teach employees how to stretch.* Employees whose jobs involve repetitive motion work such as typing can help prevent RSI by using stretching exercises. Limbering up the hands and forearms each day before starting work and again after long breaks such as the lunch hour will help eliminate the stress on muscles and tendons that can lead to RSI.

3. *Teach employees to start slowly.* Long-distance runners typically start slowly, letting their bodies adjust and their breathing find its rhythm. They pick up the pace steadily, until eventually settling in at a competitive pace. This approach is an excellent

example of how employees in jobs that are RSI prone should work. Teach employees to limber up, then begin slowly and increase their pace gradually.

4. *Avoid the use of wrist splints.* Teach employees to position their hands properly without using wrist splints. Splints can cause the muscles that they support to atrophy, thereby actually increasing the likelihood of problems.

5. *Start an exercise group.* Exercises that strengthen the hands and forearms coupled with exercises that gently stretch hand and forearm muscles can be an excellent preventive measure. Exercises that strengthen the back can help improve posture, and good posture helps prevent RSI.

6. *Select tools wisely.* RSI and other cumulative trauma disorders (CTDs) are most frequently associated with the repetitive use of VDTs and hand tools. Selecting and using hand tools properly can help prevent RSI and other CTDs. Figure 4-8 is a checklist for the proper selection and use of hand tools.

Figure 4–9 contains a summary of ergonomic strategies that can be used to help improve conditions in any organization.

Figure 4-8
Checklist for safe selection and use of hand tools.

Checklist for Safe Selection and Use of Handtools

Use Anthropometric Data

Anthropometric data has to do with human body dimensions. Such data can be used to determine the proper handle length, grip span, tool weight, and trigger length when selecting tools.

Reduce Repetition

Repetition is a hazard that can and should be reduced using such strategies as the following:

✓ Limit overtime.
✓ Change the process.
✓ Provide mechanical assists.
✓ Require breaks.
✓ Encourage stretching and strengthening exercises.
✓ Automate where possible.
✓ Rotate employees regularly.
✓ Distribute work among more employees.

Reduce the Force Required

The more force required, the more potential for damage to soft tissue. Required force can be reduced using the following strategies:

✓ Use power tools wherever possible.
✓ Use the power grip instead of the pinch grip.
✓ Spread the force over the widest possible area.
✓ Eliminate slippery, hard, and sharp gripping surfaces.
✓ Use jigs and fixtures to eliminate the pinch grip.

Minimize Awkward Postures

Awkward postures contribute to CTDs. The following strategies can reduce posture hazards:

✓ Keep the wrist in a neutral position.
✓ Keep elbows close to the body (90°–110° where bent).
✓ Avoid work that requires overhead reaching.
✓ Minimize forearm rotation.

Checklist of Ergonomic Strategies

✓ Ergonomics is the science of conforming the workplace and all of its elements to the worker.

✓ Benefits of ergonomics include the following: improved health and safety, higher morale, improved quality, improved competitiveness, decreased absenteeism, decreased tardiness, and fewer injuries.

✓ Engineering solutions are the preferred method for ergonomic hazard prevention and control. Effective engineering controls are as follows: workstation design, design of work methods, and tool design.

✓ An effective medical management program for cumulative trauma disorders is essential to the success of an employer's ergonomic program. In such a program, healthcare providers should be part of the ergonomics team.

✓ An essential element of an effective ergonomics program is training. The purpose of training is to ensure that employees are sufficiently informed about the ergonomic hazards to which they may be exposed and are thus able to participate in their own protection.

✓ Common indicators or ergonomic problems include the following: upward trend in accidents and injuries, high incidence of cumulative trauma disorders, high absenteeism, high turnover rate, employee complaints, poor quality, and too much manual lifting.

✓ Effective methods for identifying ergonomic problems include the following: general observation, questionnaires, interviews, videotaping, photography, drawing/sketching, measuring the work environment, and understanding the ergonomics of aging.

✓ Reducing the ergonomic hazards associated with VDTs can be accomplished using the following strategies:

- Arrange the keyboard properly.
- Adjust the height of the desk.
- Adjust the tilt of the keyboard.
- Encourage employees to use a soft touch on the keyboard and the mouse.
- Encourage employees to avoid wrist resting.
- Place the mouse within easy reach.
- Remove dust from mouse ball cavity.
- Locate the VDT at a proper height and distance.
- Minimize glare.
- Reduce lighting levels
- Dust the VDT screen.
- Eliminate telephone cradling.
- Require typing breaks.

✓ Repetitive strain injury (RSI) is an umbrella term that covers a number of cumulative trauma disorders. RSIs occur to the muscles, nerves, and tendons of the hands, arms, shoulders, and neck.

✓ RSI prevention strategies include the following:

- Teach employees the warning signs.
- Teach employees how to stretch.
- Teach employees to start slowly.
- Avoid the use of wrist splints.
- Start an exercise group.
- Select tools wisely.

Figure 4-9
Checklist of ergonomic strategies.

================= APPLICATION SCENARIOS =================

1. You are having a difficult time convincing your vice president that an ergonomics program is needed in your company. Develop a plan for convincing him of the benefits.
2. Assume that you have been asked to chair a committee to develop a worksite analysis program for ergonomics. You have been asked to report the committee's results to your company's executives. Develop your presentation of the program.
3. Develop an ergonomic awareness training program that can be used by any type of company.
4. Your supervisor wants to know how to predict and preclude ergonomic problems. What will you tell her?
5. Develop a plan for reducing VDT-related hazards in your company.

================= ENDNOTES =================

1. National Safety Council. *Making the Job Easier: An Ergonomics Idea Book* (Chicago: National Safety Council, 1988), pp. 1–3.
2. *Federal Register,* January 26, 1989, Vol. 54, No. 16, pp. 3904–3916.
3. OSHA 3123, 1991 (reprinted), U.S. Department of Labor, p. 1.
4. National Safety Council. *Ergonomics: A Practical Guide* (Chicago: National Safety Council, 1988), pp. 2–1 through 2–5.
5. Ibid., p. 31.
6. LaBur, Gregg. "The Age(ing) of Ergonomics," *Occupational Hazards,* April 1996, pp. 32–33.
7. Carson, Roberta. "Ergonomic Innovations: Free to a Good Company," *Occupational Hazards,* January 1996, pp. 61–64.
8. Pascarelli, Emil, and Quilter, Deborah. *Repetitive Strain Injury* (New York: John Wiley & Sons, Inc., 1994), pp. 49–62.
9. Carson, Roberta. "Ergonomically Designed Tools," *Occupational Hazards,"* September 1995. p. 50.

Stress Hazards

WORKPLACE STRESS DEFINED

Stress is a pathological, and therefore generally undesirable, human reaction to psychological, social, occupational, or environmental stimuli.[1] Stress has been defined as the reaction of the human organism to a threatening situation.[2] The stressor is an external stimuli, and stress is the response of the human body to this stimuli.

This is a clinical definition of stress. Other definitions relate more directly to workplace stress. According to Jefferson Singer of Connecticut College in New London, employers and employees disagree on how stress should be defined and what should be done to reduce it.[3]

Corporations tend to see stress as an individually based problem that is rooted in an employee's lifestyle, psychological makeup, and personality. Unions view stress as the result of excessive demands, poor supervision, or conflicting demands.[4] According to Singer, "The best definition probably includes both sets of factors."[5]

However it is defined, stress is a serious problem in the modern workplace. Stress-related medical bills and the corresponding absentee rates cost employers more than $150 billion annually. Almost 15 percent of all occupational disease claims are stress-related.

Workplace stress is the emotional state that results from a perceived difference between the demands of the job and a person's ability to cope with these demands. Since preparations and emotions are involved, workplace stress is considered a subjective state. An environment that a worker finds to be stressful may generate feelings of tension, anger, fatigue, confusion, and/or anxiety.

Workplace stress is primarily a matter of person–workload fit. The status of the person–workload fit can influence the acceptance of the work and the level of acceptable performance of that work. The perception of workload may be affected by the worker's needs and his or her level of job satisfaction. Workplace stress is further influenced by the relation between job demands and the worker's ability to meet those demands. Because workplace stress may be felt differently by different people in similar situations, it must be concluded that there are many causes of workplace stress.

SOURCES OF WORKPLACE STRESS

Sources of on-the-job stress include physical working conditions, work overload, role ambiguity, lack of feedback, personality, personal and family problems, and/or role conflict. Other sources of workplace stress are discussed in the following paragraphs.

■ Task complexity relates to the number of different demands made on the worker. A job perceived as being too complex may cause feelings of inadequacy and result in emotional stress. Repetitive and monotonous work may lack complexity so that the worker becomes bored and dissatisfied with the job and possibly experiences some stress associated with the boredom.

■ Control over the job assignment can also be a source of workplace stress. Most workers experience less stress when they participate in determining the work routine, including schedule and selection of tasks. Several studies have indicated that workers prefer to take control of their job assignment and experience less workload stress if given this opportunity.[6] A related source of stress that has been introduced in the age of high technology is from electronic monitoring. According to a study conducted at the University of Wisconsin at Madison, "Video display terminal workers who are electronically monitored suffered greater health problems than those who are not."[7]

■ A feeling of responsibility for the welfare or safety of their family members may produce on-the-job stress. Being responsible for the welfare of their families may cause workers to feel that options to take employment risks are limited. A worker may then perceive that he or she is "trapped in the job." Overly constrained employment options may lead to anxiety and stress. The feeling of being responsible for the safety of the general public has also been shown to be a stressor. Air traffic controllers are known to experience intense stress when their responsibility for public safety is tested by a near-accident event. A feeling of great responsibility associated with a job can transform a routine activity into a stress-inducing task.

■ Job security involves the risk of unemployment. A worker who believes that his or her job is in jeopardy will experience anxiety and stress. The ready availability of other rewarding employment and a feeling that one's professional skills are needed reduce the stress associated with job security issues.

■ Workload demands can stimulate stress when they are perceived as being overwhelming. These demands may involve time constraints and cognitive constraints such as speed of decision making and mandates for attention. Workload demands may also be physically overwhelming if the worker is poorly matched to the physical requirements of the job or is fatigued. Whenever the worker believes the workload to be too demanding, stress can result.

■ Psychological support from managers and co-workers gives a feeling of acceptance and belonging and helps defuse stress. A lack of such support may increase the perception of a burdensome workload and result in stress.

■ The lack of environmental safety can also be a cause of stress. Feeling that one is in danger can be a stressor. Workers need to feel safe from environmental hazards such as extreme temperatures, pressure, electricity, fire, explosives, toxic materials, ionizing radiation, noises, and dangerous machinery. To reduce the potential for stress due to environmental hazards, workers should feel that their managers are committed to safety and that their company has an effective safety program.

A poll for Northwestern National Life Insurance Company,[8] in which workers across the country were interviewed, shows an epidemic of workplace stress. Job stress sources found by this study include the following:

■ The company was recently purchased by another company.

■ Downsizing or layoffs have occurred in the past year.

- Employee benefits were significantly cut recently.
- Mandatory overtime is frequently required.
- Employees have little control over how they do their work.
- The consequences of making a mistake on the job are severe.
- Workloads vary greatly.
- Most work is machine-paced or fast-paced.
- Workers must react quickly and accurately to changing conditions.
- Personal conflicts on the job are common.
- Few opportunities for advancement are available.
- Workers cope with a great deal of bureaucracy in getting work done.
- Staffing, money, or technology is inadequate.
- Pay is below the going rate.
- Employees are rotated among shifts.

The study showed that stress may lead to decreased productivity, higher absenteeism and job turnover, poor morale, and greater numbers of stress-related illnesses.

Technological developments have also introduced new sources of stress into the workplace. The proliferation of computers is increasing stress levels in ways that were not anticipated. According to Caroline Dow and Douglas Covert of the University of Evansville, "Workplaces using computers and video display terminals (VDTs) are unwittingly breeding stress, especially among women."[9]

Dow and Covert conducted an experiment in which they observed the effect of a 16-kilohertz pure tone sound on the stress levels of college-age women. They found that anxiety and irritation were the first responses to the sound. The loudness of the sound was not a major factor; even quiet tones caused noticeable stress.[10] Figure 5–1 is a checklist of common occupational stressors.

HUMAN REACTIONS TO WORKPLACE STRESS

Human reactions to workplace stress may be grouped into the following categories: subjective or emotional (anxiety, aggression, guilt); behavioral (being prone to accidents, trembling); cognitive (inability to concentrate or make decisions); physiological (increased heart rate and blood pressure); and organizational (absenteeism and poor productivity). Continual or persistent stress has been linked to many physiological problems.[11] Initially, the effects may be psychosomatic, but with continued stress, the symptoms show up as actual organic dysfunction. The most common forms of stress-related diseases are gastrointestinal, particularly gastric or duodenal ulcers. Research has linked some autoimmune diseases with increased long-term workplace stress.[12]

The human response to workplace stress can be compared to a rubber band being stretched. As the stress continues to be applied, the rubber band stretches until a limit is reached when the rubber band breaks. For humans, various physical and psychological changes are observed with the repetitive stimuli of stress. Until the limit is reached, the harmful effects can be reversed. With an increase in intensity or duration of the stress beyond the individual's limit, the effects on the human become pathological.

M. Selye identified three stages of the human stress response: (1) alarm, (2) resistance, and (3) exhaustion.[13] The alarm reaction occurs when the stress of a threat is sensed. The stage of alarm is characterized by pallor, sweating, and an increased heart rate. This stage is usually short. It prepares the body for whatever action is necessary.

When the stress is maintained, the stage of resistance initiates a greater physical response. The alarm symptoms dissipate, and the body develops an adaptation to the stress. The capacity for adaptation during this stage is limited.

Eventually, with sustained stress, the stage of exhaustion is reached. This stage is demonstrated by the body's failure to adapt to the continued stress. Psychosomatic

Figure 5–1
Checklist of common occupational stressors.

Checklist of Common Occupational Stressors
✓ Unsafe working conditions
✓ Too much work, too little time (overload)
✓ Role ambiguity
✓ Insufficient feedback (How am I doing?)
✓ Personal problems
✓ Family problems
✓ Poor relationships with fellow employees
✓ Role conflict
✓ Complexity of the job
✓ Repetitive, monotonous tasks
✓ Lack of control
✓ Electronic monitoring
✓ Responsibility for supporting a family
✓ Insufficient employment options
✓ Responsibility for the performance of others
✓ Responsibility for the safety of others
✓ Job security fears
✓ Lack of acceptance from fellow employees
✓ Corporate buyouts and mergers
✓ Downsizing and layoffs
✓ Wage and benefit disputes
✓ Fear of the consequence of mistakes
✓ Performance appraisals
✓ Pace of the work (too fast or too slow)
✓ Change
✓ Technology

diseases such as gastric ulcers, colitis, rashes, and autoimmune disorders may begin during this stage. The tendency to develop a specific stress-related disease may be partially predetermined by heredity, personal habits such as smoking, and personality.

From an evolutionary viewpoint, the adverse effects of stress on health may be considered to be a maladaptation of humans to stress. What does this tell us? Either we (1) learn to do away with all stress (unlikely); (2) avoid all stressful situations (equally unlikely); (3) learn to adapt to being sick because of stress (undesirable); or (4) learn to adapt to workplace stress (the optimal choice). The first step in learning to adapt to stress is understanding the amount of stress to which we are subjected. Figure 5–2 is a checklist of human reactions to workplace stress.

REDUCING WORKPLACE STRESS

Not all sources of stress on the job can be eliminated, and employment screening is unlikely to identify all those who are sensitive to stress. People can learn to adapt to stress, however. Training can help people recognize and deal with stress effectively. Employees need to know what is expected of them at any given time and to receive recognition when it is deserved. Managers can reduce role ambiguity and stress caused by lack of feedback by providing frequent feedback.

Figure 5–2
Checklist of human reactions to workplace stress.

Checklist of Human Reactions to Workplace Stress

- ✓ Anxiety
- ✓ Aggression
- ✓ Guilt
- ✓ Becoming accident prone
- ✓ Inability to concentrate
- ✓ Inability to make decisions
- ✓ Poor decisions
- ✓ Increased blood pressure
- ✓ Increased lateness
- ✓ Absenteeism
- ✓ Gastrointestinal disease
- ✓ Autoimmune disease
- ✓ Fatigue
- ✓ Sluggishness

Stress can result from low participation or lack of job autonomy. A manager can help employees realize their full potential by helping them match their career goals with the company's goals and giving them more control over their jobs.

Managers can help design jobs in ways that lead to worker satisfaction, thereby lessening work stress. Physical stress can be reduced by improving the work environment and establishing a sound safety and health program. Managers can also assist in the effort to provide varied and independent work with good possibilities for contact and collaboration with fellow workers and for personal development.

Organizational approaches to coping with work stress include avoiding a monotonous, mechanically controlled pace, standardized motion patterns, and constant repetition of short-cycle operations. Other stress-inducing work design features to avoid include jobs that do not make use of a worker's knowledge and initiative, that lack human contact, and that have authoritarian-type supervision.

There are also several individual approaches to coping with stress. One of the most important factors in dealing with stress is learning to recognize its symptoms and taking them seriously. Handling stress effectively should be a lifelong activity that gets easier with practice. Keeping a positive mental attitude can help defuse some otherwise stressful situations.

People can analyze stress-producing situations and decide what is worth worrying about. Individuals can effectively respond to a stressful workload by delegating responsibility instead of carrying the entire load. Relaxation techniques can also help reduce the effects of stress. Some common relaxation methods include meditation, biofeedback, music, and exercise.

The Northwestern National Life Insurance Company recommends the following strategies for workplace stress reduction:[14]

- Management recognizes workplace stress and takes steps regularly to reduce this stress.
- Mental health benefits are provided in the employee's health insurance coverage.
- The employer has a formal employee communications program.
- Employees are given information on how to cope with stress.
- Workers have current, accurate, and clear job descriptions.

- Management and employees talk openly with one another.
- Employees are free to talk with each other during work.
- Employers offer exercise and other stress reduction classes.
- Employees are recognized and rewarded for their contributions.
- Work rules are published and are the same for everyone.
- Child-care programs are available.
- Employees can work flexible hours.
- Perks are granted fairly based on a person's level in the organization.
- Workers have the training and technology access that they need.
- Employers encourage work and personal support groups.
- Workers have a place and time to relax during the work day.
- Elder-care programs are available.
- Employees' work spaces are not crowded.
- Workers can put up personal items in their work areas.
- Management appreciates humor in the workplace.

In addition to the strategies recommended by the Northwestern National Life Insurance Company, several others have been identified. Writing for *Occupational Hazards*, S. L. Smith recommends the following ways to reduce stress in the workplace:[15]

- Match workload and pace to the training and abilities of employees.
- Make an effort to match work schedules with the personal lives of employees.
- Clearly define work roles.
- Before giving employees additional duties beyond their normal work roles, make sure they receive the necessary training.
- Promote teamwork among employees and encourage it throughout the organization.
- Involve employees in making decisions that affect them.
- Inform employees in a timely manner of organizational changes that might affect them.

There is no one clear answer to workplace stress. The suggestions given here are a good starting place for management and employees to begin the process of being aware of and dealing effectively with workplace stress.

STRESS AND WORKERS' COMPENSATION

There are serious problems with workers' compensation in the United States. On the one hand, there is evidence of abuse of the system. On the other hand, many injured workers who are legitimately collecting benefits suffer a substantial loss of income. Complaints about workers' compensation are common from all parties involved with the system (employers, employees, and insurance companies).

An example of how the workers' compensation system can be abused is an overweight deputy sheriff who applied for benefits due to stress. He was distraught over the breakup of his extramarital love affair and a poor performance evaluation. This individual is just one of thousands who are claiming that job stress has disabled them to the point that workers' compensation is justified. In 1980, there were so few stress claims that they were not even recorded as a separate category. By now, they represent a major and costly category.

Stress claims are more burdensome than physical claims because they are typically reviewed in an adversarial environment. This leads to the involvement of expert medical witnesses and attorneys. As a result, even though the benefits awarded for stress-related injuries are typically less than those awarded for physical injuries, the cost of stress claims is often higher because of litigation.

═════════ **APPLICATION SCENARIOS** ═════════

1. You've been asked by the CEO of your company to attend a board of directors meeting and give a brief explanation of workplace stress and why the company should be concerned about it. Write out your presentation.

2. Using the checklist in this chapter, assess an organization to identify potential sources of stress. The organization may be your current employer, a past employer, or any other organization with which you are familiar.

3. Based on the assessment you completed in Scenario 2, develop a plan for reducing workplace stress in that organization.

═════════ **ENDNOTES** ═════════

1. Fraser, T. M. *The Worker at Work* (New York: Taylor & Francis, 1989), p. 103.

2. Grandjean, E. *Fitting the Task to the Man, A Textbook of Occupational Ergonomics*, 4th ed. (New York: Taylor & Francis, 1988), pp. 175–176.

3. "Causes and Cures for Stress Still Misunderstood," *Occupational Health & Safety Letter*, November 28, 1990, Vol. 20, No. 24, p. 196.

4. Ibid.

5. Ibid.

6. Grandjean, E. *Fitting the Task to the Man,* pp. 176–177.

7. "Electronic Monitoring Causes Worker Stress," *Occupational Health & Safety Letter,* October 17, 1990, Vol. 20, No. 21, p. 168.

8. Cope, L. "Quiz Developed to Determine Workplace Stress," *Tallahassee Democrat*, June 2, 1991, pp. 1E, 3E.

9. "Computer Sound in Workplace Linked to Stress," *Occupational Health & Safety Letter*, August 22, 1990, Vol. 20, No. 17, p. 141.

10. Ibid.

11. Fraser, T. M. *The Worker at Work,* pp. 104, 110–111.

12. Konz, S. *Work Design: Industrial Ergonomics* (Worthington, OH: Publishing Horizons, 1990), pp. 347–390.

13. Fraser, T. M. *The Worker at Work*, pp. 111–115.

14. Cope, L. "Quiz Developed to Determine Workplace Stress."

15. Smith, S. L. "Combating Stress," *Occupational Hazards*, March 1994, p. 57.

CHAPTER SIX

Mechanical Hazards and Machine Safeguarding

Mechanical hazards are those associated with power-driven machines, whether automated or manually operated. Concerns about mechanical hazards date back to the Industrial Revolution and the earliest days of mechanization. Machines driven by steam, hydraulic, and/or electric power introduced new hazards into the workplace. In spite of advances in safeguarding technologies and techniques, mechanical hazards are still a major concern today. In addition, automated machines have introduced new concerns.

COMMON MECHANICAL INJURIES

In an industrial setting, people interact with machines that are designed to drill, cut, shear, punch, chip, staple, stitch, abrade, shape, stamp, and slit such materials as metals, composites, plastics, and elastomers. If appropriate safeguards are not in place or if workers fail to follow safety precautions, these machines can apply the same procedures to humans. When this happens, the types of mechanical injuries that result are typically the result of cutting, tearing, shearing, crushing, breaking, straining, or puncturing (see Figure 6–1). Information about each of these hazards is provided in the following paragraphs.

Cutting and Tearing

A cut occurs when a body part comes in contact with a sharp edge. The human body's outer layer is arranged from the outside in as follows: *epidermis*, which is the tough outer covering of the skin; *dermis*, which constitutes the greatest part of the skin's thickness; *capillaries*, which are tiny blood vessels that branch off the small arteries and veins in the dermis; *veins*, which are blood vessels that collect blood from the capillaries and return it

Figure 6-1
Checklist of common mechanical injuries.

Checklist of Common Mechanical Injuries
- ✓ Cutting
- ✓ Tearing
- ✓ Shearing
- ✓ Crushing
- ✓ Breaking
- ✓ Straining
- ✓ Puncturing

to the heart; and *arteries,* which are the larger vessels that carry blood from the heart to the capillaries in the skin. The seriousness of cutting or tearing the skin depends on how much damage is done to the skin, veins, arteries, muscles, and even bones.

Shearing

To understand what shearing is, think of a paper cutter. It shears the paper. Power-driven shears for severing paper, metal, plastic, elastomers, and composite materials are widely used in manufacturing. In times past, such machines often amputated fingers and hands. Such tragedies would typically occur when operators reached under the shearing blade to make an adjustment or to place materials there and activate the blade before fully removing their hand.

Crushing

Injuries from crushing can be particularly debilitating, painful, and difficult to heal. They occur when a part of the body is caught between two hard surfaces that progressively move together, thereby crushing anything between them. Crushing hazards can be divided into two categories: *squeeze-point* types and *run-in points*.

Squeeze-point hazards exist where two hard surfaces, at least one of which must be in motion, push close enough together to crush any object that may be between them. The process can be slow, as in a manually operated vice, or fast, as with a metal-stamping machine.

Run-in point hazards exist where two objects, at least one of which is rotating, come progressively closer together. Any gap between them need not become completely closed. It need only be smaller than the object or body part lodged in it. Meshing gears and belt pulleys are examples of run-in point hazards.

Body parts can also be crushed in other ways, for example, a heavy object falling on a foot or a hammer hitting a finger.

Breaking

Machines used to deform engineering materials in a variety of ways can also cause broken bones. A break in a bone is known as a fracture. Fractures are classified as simple, compound, complete, and incomplete.

A simple fracture is a break in a bone that does not pierce the skin. A compound fracture is a break that has broken through the surrounding tissue and skin. A complete fracture divides the affected bone into two or more separate pieces. An incomplete fracture leaves the affected bone in one piece but cracked.

Fractures are also classified as transverse, oblique, and comminuted. A transverse fracture is a break straight across the bone. An oblique fracture is diagonal. A comminuted fracture exists when the bone is broken into a number of small pieces at the point of fracture.

Straining and Spraining

There are numerous situations in a work setting when straining of muscles or spraining of ligaments is possible. A strain results when muscles are overstretched or torn. A sprain is the result of torn ligaments in a joint. Strains and sprains can cause swelling and intense pain.

Puncturing

Punching machines that have sharp tools can puncture a body part if safety precautions are not observed or if appropriate safeguards are not in place. Puncturing results when an object penetrates straight into the body and pulls straight out, creating a wound in the shape of the penetrating object. The greatest hazard with puncture wounds is the potential for damage to internal organs.

SAFEGUARDING DEFINED

All of the hazards explained in the previous section can be reduced by the application of appropriate safeguards. C.F.R. 1910 Subpart O contains the OSHA standards for machinery and machine guarding (1910.211–1910.222). The National Safety Council defines safeguarding as follows:

> . . . machine safeguarding is to minimize the risk of accidents of machine–operator contact. The contact can be:
>
> 1. An individual making the contact with the machine—usually the moving part—because of inattention caused by fatigue, distraction, curiosity, or deliberate chance taking;
> 2. From the machine via flying metal chips, chemical and hot metal splashes, and circular saw kickbacks, to name a few;
> 3. Caused by the direct result of a machine malfunction, including mechanical and electrical failure.[1]

Safeguards can be broadly categorized as point-of-operation guards, point-of-operation devices, and feeding/ejection methods. The various types of safeguards in these categories are explained later in this chapter.

REQUIREMENTS FOR ALL SAFEGUARDS

The various machine motions present in modern industry involve mechanisms that rotate, reciprocate, or do both. This equipment includes tools, bits, chucks, blades, spokes, screws, gears, shafts, belts, and a variety of different types of stock. Safeguards can be devised to protect workers from harmful contact with such mechanisms while at the same time allowing work to progress at a productive rate. The National Safety Council has established the following requirements for safeguards.[2]

1. *Prevent contact.* Safeguards should prevent human contact with any potentially harmful machine part. The prevention extends to machine operators and any other person who might come in contact with the hazard.

2. *Be secure and durable.* Safeguards should be attached so that they are secure. This means that workers cannot render them ineffective by tampering with or disabling them. This is critical because removing safeguards in an attempt to speed production is a common practice. Safeguards must also be durable enough to withstand the rigors of the workplace. Worn-out safeguards won't protect workers properly.

3. *Protect against falling objects.* Objects falling onto moving machine mechanisms increase the risk of accidents, property damage, and injury. Objects that fall on a moving part can be quickly hurled out, creating a dangerous projectile. Therefore, safeguards must do more than just prevent human contact. They must also shield the moving parts of machines from falling objects.

4. *Create no new hazard.* Safeguards should overcome the hazards in question without creating new ones. For example, a safeguard with a sharp edge, unfinished surface, or protruding bolts introduces new hazards while protecting against the old.

5. *Create no interference.* Safeguards can interfere with the progress of work if they are not properly designed. Such safeguards are likely to be disregarded or disabled by workers feeling the pressure of production deadlines.

6. *Allow safe maintenance.* Safeguards should be designed to allow the more frequently performed maintenance tasks (e.g., lubrication) to be accomplished without the removal of guards. For example, locating the oil reservoir outside the guard with a line running to the lubrication point will allow for daily maintenance without removing the guard.

Design and construction of safeguards are highly specialized activities requiring a strong working knowledge of machines, production techniques, and safety. However, it is critical that all of the factors explained in this section be considered and accommodated during the design process.

POINT-OF-OPERATION GUARDS[3]

Guards are most effective when used at the point of operation, which is where hazards to humans exist. Point-of-operation hazards are those caused by the shearing, cutting, or bending motions of a machine. Pinch-point hazards result from guiding material into a machine or transferring motion (e.g., from gears, pressure rollers, or chains and sprockets). Single-purpose safeguards, since they guard against only one hazard, typically are permanently fixed and nonadjustable. Multiple-purpose safeguards, since they guard against more than one hazard, typically are adjustable.

Point-of-operation guards are of three types, each with its own advantages and limitations: fixed, interlocked, and adjustable.

- Fixed guards provide a permanent barrier between workers and the point of operation. They offer the following advantages: They are suitable for many specific applications, can be constructed in-plant, require little maintenance and are suitable for high-production, repetitive operations. Limitations include the following: They sometimes limit visibility, are often limited to specific operations, and sometimes inhibit normal cleaning and maintenance.

- Interlocked guards shut down the machine when the guard is not securely in place or is disengaged. The main advantage of this type of guard is that it allows safe access to the machine for removing jams or conducting routine maintenance without the need for taking off the guard. There are also limitations. Interlocked guards require careful adjustment and maintenance and, in some cases, can be easily disengaged.

- Adjustable guards provide a barrier against a variety of different hazards associated with different production operations. They have the advantage of flexibility. However, they do not provide as dependable a barrier as other guards do, and they require frequent maintenance and careful adjustment.

POINT-OF-OPERATION DEVICES[4]

A number of different point-of-operation devices can be used to protect workers. The most widely used are explained in the following paragraphs.

■ Photoelectric devices are optical devices that shut down the machine whenever the light field is broken. These devices allow operators relatively free movement. They do have limitations, including the following: They do not protect against mechanical failure, they require frequent calibration, and they can be used only with machines that can be stopped.

■ Radio-frequency devices are capacitance devices that brake the machine if the capacitance field is interrupted by a worker's body or another object. These devices have the same limitations as photoelectric devices.

■ Electromechanical devices are contact bars that allow only a specified amount of movement between the worker and the hazard. If the worker moves the contact bar beyond the specified point, the machine will not cycle. These devices have the limitation of requiring frequent maintenance and careful adjustment.

■ Pullback devices pull the operator's hands out of the danger zone when the machine starts to cycle. These devices eliminate the need for auxiliary barriers. However, they also have limitations. They limit operator movement, must be adjusted for each individual operator, and require close supervision to ensure proper use.

■ Restraint devices hold the operator back from the danger zone. They work well, with little risk of mechanical failure. However, they do limit the operator's movement, must be adjusted for each individual operator, and require close supervision to ensure proper use.

■ Safety trip devices include trip wires, trip rods, and body bars. All of these devices stop the machine when tripped. They have the advantage of simplicity. However, they are limited in that all controls must be activated manually. They protect only the operator and may require the machine to be fitted with special fixtures for holding work.

■ Two-hand controls require the operator to use both hands concurrently to activate the machine (e.g., a paper-cutter or metal-shearing machine). This ensures that hands cannot stray into the danger zone. Although these controls do an excellent job of protecting the operator, they do not protect onlookers or passers-by. In addition, some two-hand controls can be tampered with and made operable using only one hand.

■ Gates provide a barrier between the danger zone and workers. Although they are effective at protecting operators from machine hazards, they can obscure the work, making it difficult for the operator to see.

FEEDING AND EJECTION SYSTEMS[5]

Feeding and ejection systems can be effective safeguards if properly designed and used. The various types of feeding and ejection systems available for use with modern industrial machines are summarized in the following paragraphs.

■ Automatic feed systems feed stock to the machine from rolls. Automatic feeds eliminate the need for operators to enter the danger zone. Such systems are limited in the types and variations of stock that they can feed. They also typically require an auxiliary barrier guard and frequent maintenance.

■ Semiautomatic feed systems use a variety of approaches for feeding stock to the machine. Prominent among these are chutes, moveable dies, dial feeds, plungers, and sliding bolsters. They have the same advantages and limitations as automatic feed systems.

■ Automatic ejection systems eject the work pneumatically or mechanically. The advantage of either approach is that operators don't have to reach into the danger zone to retrieve workpieces. However, these systems are restricted to use with relatively small stock. Potential hazards include blown chips or debris and noise. Pneumatic ejectors can be quite loud.

■ Semiautomatic ejection systems eject the work using mechanisms that are activated by the operator. Consequently, the operator does not have to reach into the danger zone to retrieve workpieces. These systems do require auxiliary barriers and can be used with a limited variety of stock.

ROBOT SAFEGUARDS

Robots have become commonplace in modern industry. Only the guarding aspects of robot safety are covered in this section. The main hazards associated with robots are (1) entrapment of a worker between a robot and a solid surface. (2) impact with a moving robot arm, and (3) impact with objects ejected or dropped by the robot.

The best guard against these hazards is to erect a physical barrier around the entire perimeter of a robot's work envelope (the three-dimensional area established by the robot's full range of motion). This physical barrier should be able to withstand the force of the heaviest object that a robot could eject.

Various types of shutdown guards can also be used. A guard containing a sensing device that automatically shuts down the robot if any person or object enters its work envelope can be effective. Another approach is to put sensitized doors or gates in the perimeter barrier that automatically shut down the robot as soon as they are opened.

These types of safeguards are especially important because robots can be deceptive. A robot that is not moving at the moment may simply be at a stage between cycles. Without warning, it might make sudden and rapid movements that could endanger any person inside the work envelope.

LOCKOUT/TAGOUT SYSTEMS

One of the most effective safeguarding approaches in use today is the lockout/tagout system. This method was especially designed to protect against the unexpected startup of a machine that is supposed to be turned off. This is important because OSHA statistics show that six percent of all workplace fatalities are caused by the unexpected activation of machines while they are being serviced, cleaned, or otherwise maintained.[6] OSHA's lockout/tagout standard is C.F.R. 1910.147 and 1910.331.

In a lockout system, a padlock is placed through a gate covering the activating mechanism or is applied in some other manner to prevent a machine from being turned on until the lock is removed. The lock usually has a label that gives the name, department, and telephone extension of the person who put it on. It may also carry a message such as the following: *"This lock is to be removed only by _____ ."* The name in the blank is the person who applied the lock.

A tagout system is exactly like a lockout system except a tag is substituted for the lock. Tags should be used only in cases where a lock is not feasible. Sometimes, tags and locks may be used together.

The following examples demonstrate why lockout/tagout systems are so important: (1) An employee is cleaning the guarded side of an operating granite saw when he is pulled into it and killed; (2) an employee is cleaning scrap from under a shear when another employee hits the activation button and decapitates him; and (3) an employee is cleaning paper from a waste crusher when he falls into it and is crushed.[7]

Accidents such as these were the driving force behind the development of OSHA Standard 29 C.F.R. 1910.147 and its follow-on Instruction STD 1–7.3. The standard covers all general industry in the United States and requires the use of lockout or tagout systems to protect against the unexpected activation of machines and equipment. Instruction STD 1–7.3 clarifies the standard to require the use of lockout systems wherever possible. Tagout systems are to be used only in those instances where lockout systems are not feasible. "Instruction STD 1–7.3 states that when an employer uses tags instead of locks, they must demonstrate that the tagout program will provide a level of safety equivalent to a lockout program."[8] It is estimated by OSHA that full compliance with the lockout/tagout standard will prevent 120 accidental deaths, 29,000 serious injuries, and 32,000 minor injuries every year.[9]

OSHA identified the following four factors that are essential for effective lockout/tagout systems:

1. *Attention to detail.* It is important to have procedures covering all steps and allowing for all contingencies.
2. *Extensive training.* Employers must teach workers about lockout/tagout procedures and why it is important to use them.
3. *Reinforcement of training.* Employers must reinforce the training provided by stressing the importance of safety principles and practices.
4. *Disciplinary action.* Violators of lockout/tagout procedures and other general safety principles and practices must be disciplined.[10]

GENERAL PRECAUTIONS

The types of safeguards explained in this chapter are critical. In addition to these specific safeguards, there are also a number of general precautions that apply across the board in settings where machines are used. Some of the more important general precautions are as follows:

- All operators should be trained in the safe operation and maintenance of their machines.
- All machine operators should be trained in the emergency procedures to take when accidents occur.
- All employees should know how to activate emergency shutdown controls. This means knowing where the controls are and how to activate them.
- Inspection, maintenance, adjustment, repair, and calibration of safeguards should be carried out regularly.
- Supervisors should ensure that safeguards are properly in place when machines are in use. Employees who disable or remove safeguards should be disciplined appropriately.
- Operator teams (two or more operators) of the same system should be trained in coordination techniques and proper use of devices that prevent premature activation by a team member.
- Operators should be trained and supervised to ensure that they dress properly for the job. Long hair, loose clothing, neckties, rings, watches, necklaces, chains, and earrings can become caught in equipment and, in turn, pull the employee into the hazard zone.
- Shortcuts that violate safety principles and practices should be avoided. The pressures of deadlines should never be the cause of unsafe work practices.
- Other employees who work around machines but do not operate them should be made aware of the emergency procedures to take when an accident occurs.

BASIC PROGRAM CONTENT

Machine safeguarding should be organized, systematic, and comprehensive. A company's safeguarding program should have at least the following elements:

- Safeguarding policy that is part of a broader company-wide safety and health policy
- Machine/hazard analysis
- Lockout/tagout (materials and procedures)
- Employee training

Checklist of Problems and Corresponding Actions	
Problem	**Action**
Machine operating without the safety guard.	Stop machine immediately and activate the safety guard.
Maintenance worker cleaning a machine that is operating.	Stop machine immediately and lock or tag it out.
Visitor to the shop is wearing a necktie as he observes a lathe in operation.	Immediately pull the visitor back and have him remove the tie.
An operator is observed disabling a guard.	Stop the operator, secure the guard, and take disciplinary action.
A robot is operating without a protective barrier.	Stop the robot and erect a barrier immediately.
A machine guard has a sharp, ragged edge.	Stop the machine and eliminate the sharp edge and ragged burrs by rounding it off.

Figure 6–2
Selected examples of problems and corresponding actions.

- Comprehensive documentation
- Periodic safeguarding audits (at least annually)

TAKING CORRECTIVE ACTION

What should be done when a mechanical hazard is observed? The only acceptable answer to this question is, take *immediate corrective action*. The specific action indicated will depend on what the problem is. Figure 6–2 shows selected examples of problems and corresponding corrective actions. Figure 6-3 is a checklist for identifying mechanical hazards.

These are only a few of the many different types of problems that require corresponding corrective action. Regardless of the type of problem, the key to responding is immediacy. Waiting to take corrective action can be fatal.

Checklist for Identifying Mechanical Hazards

✓ Are point-of-operation guards being properly used where appropriate?

✓ Are point-of-operation devices being properly used where appropriate?

✓ Are the safest feeding/ejection methods being employed?

✓ Are lockout/tagout systems being employed as appropriate?

✓ Are automatic shutdown systems in place, clearly visible, and accessible?

✓ Are safeguards properly maintained, adjusted, and calibrated as appropriate?

Figure 6–3
Mechanical hazards checklist.

APPLICATION SCENARIOS

1. Assume you are the new safety director for Precision Machining Company (PMC). PMC uses the following types of machining processes: milling, turning, and shearing. What types of injuries might occur at PMC?
2. A sales representative is trying to sell you a new safeguard for your company's machine tools. What criteria can you apply to determine the effectiveness of the safeguard?
3. The machines in your company's shop have no safeguards. You have convinced the shop foreman to install point-of-operation guards. He wants to make sure that operators don't become frustrated and take the safeguards off when performing routine maintenance. What type of guards should you recommend?
4. Your company had a tagout program but stopped using it because machine operators continually ignored the tags. Outline and explain the specific steps you would recommend for ensuring an effective tagout system.
5. Develop a plan for a comprehensive machine safeguarding program at Bailey Brothers Machining.

ENDNOTES

1. National Safety Council. *Guards: Safeguarding Concepts Illustrated*, 5th ed. (Chicago: National Safety Council, 1987), p. 1.
2. National Safety Council. *Guards: Safeguarding Concepts Illustrated*, pp. 2–3.
3. Ibid., p. 36.
4. Ibid., pp. 38–39.
5. Ibid., p. 44.
6. Caruey, A. "Lock Out the Chance for Injury," *Safety & Health*, May 1991, Vol. 143, No. 5, p. 46.
7. Ibid.
8. Ibid.
9. Ibid.
10. Ibid.

CHAPTER SEVEN

Falling and Lifting Hazards with Eye, Head, and Foot Protection

Some of the most common accidents in the workplace happen as the result of slipping, falling, and improper lifting. Impact from a falling object is also a common cause of accidents. This chapter provides the information needed by modern safety and health professionals to prevent such accidents.

CAUSES OF FALLS

In a typical year, more than 10,000 workers will lose their lives in falls. More than 16 percent of all disabling work-related injuries are the result of falls.[1] Clearly, falls are a major concern of safety and health professionals. According to Kohr, the primary causes of falls are

1. A foreign object on the walking surface
2. A design flaw in the walking surface
3. Slippery surfaces
4. An individual's impaired physical condition[2]

A foreign object is any object that is out of place or in a position to trip someone or to cause a slip. There are an almost limitless number of design flaws that might cause a fall. A poorly designed floor covering, a ladder that does not seat properly, or a catwalk that gives way are all examples of design flaws that might cause falls. Slippery surfaces are particularly prevalent in industrial plants where lubricants and cleaning solvents are used.

Automobile accidents are often caused when a driver's attention is temporarily drawn away from the road by a visual distraction. This is also true in the workplace. Anything that distracts workers visually can cause a fall. When a person's physical condition is impaired for any reason, the potential for falls increases. This is a particularly common problem among aging workers. Understanding these causes is the first step in developing fall prevention techniques.

KINDS OF FALLS

Falling from ladders and other elevated situations is covered later in this chapter. This section deals with the more common surface falls. According to Miller, such falls can be divided into the following four categories:

- Trip and fall accidents occur when workers encounter an unseen foreign object in their path. When the employees' foot strikes the object, he or she trips and falls.
- Stump and fall accidents occur when a worker's foot suddenly meets a sticky surface or a defect in the walking surface. Expecting to continue at the established pace, the worker falls when his or her foot is unable to respond properly.
- Step and fall accidents occur when a person's foot encounters an unexpected step down (e.g., a hole in the floor or a floorboard that gives way). This can also happen when an employee thinks he or she has reached the bottom of the stairs when, in reality, there is one more step.
- Slip and fall accidents occur when the worker's center of gravity is suddenly thrown out of balance (e.g., an oily spot causes a foot to shoot out from under the worker).[3]

Miller summarizes his finding relating to surface falls:

> The most common is the slip and fall accident. Foot contact is broken, and the individual attempts to right himself. A recovery of equilibrium is reflexive and not under conscious control in most cases. If the pedestrian strikes the walking surface with a fleshy part of the body, the injuries are likely to be minimal. If, on the other hand, the victim strikes a bony body part in the fall, the injuries are likely to be more severe.[4]

WALKING AND SLIPPING

Judging by the number of injuries that occur each year as the result of slipping, it is clear that walking can be hazardous to a worker's health. This is, in fact, the case when walking on an unstable platform. Goldsmith defines a stable platform for walking as "a walking surface with a high degree of traction, free from obstructions and therefore safe."[5] It follows that an unstable platform is one lacking traction, or one on which there are obstructions, or both.

Strategies for Preventing Slips

Modern safety and health professionals are concerned with preventing slips and falls. Slip prevention should be a part of the company's larger safety and health program. Here are some strategies that can be used to help prevent slipping:

1. *Choose the right material from the outset.* Where the walking surface is to be newly constructed or an existing surface is to be replaced, safety and health professionals should encourage the selection of surface materials that have the highest possible coefficient of friction. Getting it right from the start is the best way to prevent slipping accidents.

2. *Retrofit an existing surface.* If it is too disruptive or too expensive to replace a slippery surface completely, retrofit it with friction enhancement devices or materials. Such devices/materials include runners, skid strips, carpet, grooves, abrasive coatings, grills, and textured coverings.

3. *Practice good housekeeping.* Regardless of the type of surface, keep it clean and dry. Spilled water, grease, oil, solvents, and other liquids should be removed immediately. When the surface is wet intentionally, as when cleaning or mopping, rope off the area and erect warning signs.

4. *Require nonskid footwear.* Employees who work in areas where slipping is likely to be a problem should be required to wear shoes with special nonskid soles. This is no different from requiring steel-toed boots to protect against falling objects. Nonskid footwear should be a normal part of a worker's personal protective equipment.

5. *Inspect surfaces frequently.* Employees who are working to meet production deadlines may be so distracted that they don't notice a wet surface, or they may notice it but feel too rushed to do anything about it. Consequently, safety and health professionals should conduct frequent inspections and act immediately when a hazard is identified.[6]

SLIP AND FALL PREVENTION PROGRAMS[7]

A company's overall safety and health program should include a slip and fall prevention component. According to Kohr, such a component should have the following elements:

1. *A policy statement/commitment.* Statement to convey management's commitment. Areas that should be included in the policy statement are management's intent, scope of activity, responsibility, accountability, the safety professional's role, authority, and standards.

2. *Review and acceptance of walkways.* Contain the criteria that will be used for reviewing all walking surfaces and determining if they are acceptable. For example, a criterion might be a minimum coefficient of friction value. Regardless of the criteria, the methodology that will be used for applying them to the review and acceptance of walkways should also be explained.

3. *Reconditioning and retrofitting.* Include recommendations and timetables for reconditioning or retrofitting existing walking surfaces that do not meet review and acceptance criteria.

4. *Maintenance standards and procedures.* State the maintenance standards for walking surfaces (e.g., How often should surfaces be cleaned, resurfaced, replaced, etc.?). In addition, this section should contain procedures for meeting the standards.

5. *Inspection, audits, tests, and records.* Provide a comprehensive list of inspections, audits, and tests (including the types of tests) that will be done, how frequently, and where. Give records of the results.

6. *Employee footwear program.* Specify the type of footwear required of employees who work on different types of walking surfaces.

7. *Defense methods for legal claims.* Outline the company's legal defenses so that aggressive action can be taken immediately should a lawsuit be filed against the company. In such cases, it is important to be able to show that the company has not been negligent (e.g., the company has a slip and fall prevention program that is in effect).

8. *Measurement of results.* Contain the following two parts: (1) an explanation of how the program will be evaluated and how often (e.g., comparison of yearly, quarterly, or monthly slip and fall data); (2) records of the results of these evaluations.

OSHA FALL PROTECTION STANDARDS

The OSHAct mentions fall protection in several places. Although the General Industry Standards are silent on fall protection, the problem is covered in the following subparts:

Subpart D	Walking/working surfaces
Subpart F	Powered platforms, manlifts, and vehicle-mounted work platforms
Subpart R	Special industries

In addition to these OSHA standards, the American National Standards Institute (ANSI) publishes a Fall Protection Standard (ANSI Z359.1: *Safety Requirements for Personal Fall Arrest Systems, Subsystems, and Components*). The most comprehensive and most controversial fall protection standard is OSHA's Fall Protection Standard for the construction industry (Subpart M of 29 C.F.R. 1926).

OSHA's Fall Protection Standard for Construction

OSHA's current Fall Protection Standard sets the *trigger height* at six feet. This means that any construction employee working higher than six feet off the ground must use a fall protection device such as a safety harness and line.

This trigger height means that virtually every small residential builder and roofing contractor is subject to the standard. Because most residential builders and roofing contractors are small and fall protection equipment is expensive, Subpart M of 29 C.F.R. 1926 is a source of much controversy.

OSHA officials argue that the six-foot trigger height saves up to 80 lives per year and prevents more than 56,000 injuries. The rationale is that six percent of all lost-time fall injuries in the construction industry are caused by falls from less than 10 feet. Opponents counter that the cost of complying with the standard is almost $300 million annually. Commercial contractors, whose employees typically work much higher than the 6- to 16-foot range are not concerned about the height controversy. Consequently, OSHA is under intense pressure to waive the six-foot trigger height for residential builders and roofers.

LADDER SAFETY

Jobs that involve the use of ladders introduce their own set of safety problems, one of which is an increased potential for falls. The National Safety Council recommends that ladders be inspected before every use and that employees who use them follow a set of standard rules.[8]

Inspecting Ladders

Taking a few moments to look over a ladder carefully before using it can prevent a fall. The National Safety Council recommends the following when inspecting a ladder:

- See if the ladder has the manufacturer's instruction label on it.
- Determine whether the ladder is strong enough.
- Read the label specifications about weight capacity and applications.
- Look for the following conditions: cracks on side rails; loose rungs, rails, or braces; damaged connections between rungs and rails.
- Check for heat damage and corrosion.
- Check wooden ladders for moisture that might cause them to conduct electricity.
- Check metal ladders for burrs and sharp edges.
- Check fiberglass ladders for signs of blooming (deterioration of exposed fiberglass).[9]

Dos and Don'ts of Ladder Use

Many accidents involving ladders result from improper use. Following a simple set of rules for the proper use of ladders can reduce the risk of falls and other ladder-related accidents. The National Safety Council recommends the following dos and don'ts of ladder use:

- Check for slipperiness on shoes and ladder rungs.
- Limit a ladder to one person at a time.
- Secure the ladder firmly at the top and bottom.
- Set the ladder's base on a firm, level surface.
- Apply the four-to-one ratio (base one foot away from the wall for every four feet between the base and the support point).
- Face the ladder when climbing up or down.
- Barricade the base of the ladder when working near an entrance.
- Don't lean a ladder against a fragile, slippery, or unstable surface.
- Don't lean too far to either side while working (stop and move the ladder).
- Don't rig a makeshift ladder; use the real thing.
- Don't allow more than one person at a time on a ladder.
- Don't allow your waist to go any higher than the last rung when reaching upward on a ladder.
- Don't carry tools in your hands while climbing a ladder.
- Don't place a ladder on a box, table, or bench to make it reach higher.[10]

OSHA standards for walking and working surfaces and ladder safety are set forth in 29 C.F.R. Part 1910 (Subpart D). The standards contained in Subpart D are as follows:

1910.21	Definitions
1910.22	General requirements
1910.23	Guarding floor and wall openings and holes
1910.24	Fixed industrial stairs
1910.25	Portable wood ladders
1910.26	Portable metal ladders
1910.27	Fixed ladders
1910.28	Safety requirements for scaffolding
1910.29	Manually propelled mobile ladder stands and scaffolds (towers)
1910.30	Other working surfaces
1910.31	Sources of standards
1910.32	Standards organizations

IMPACT AND ACCELERATION HAZARDS

An employee working on a catwalk drops a wrench. The falling wrench accelerates over the 20-foot drop and strikes an employee below. Had the victim not been wearing a hard hat he might have sustained serious injuries from the impact. A robot loses its grip on a part, slinging it across the plant and striking an employee. The impact from the part breaks one of the employee's ribs. These are examples of accidents involving acceleration and impact. So is any type of fall since, having fallen, a person's rate of fall accelerates until striking a surface (impact). Motor vehicle accidents are also acceleration/impact instances.

Since falls were covered in the previous section, this section will focus on hazards relating to the acceleration and impact of objects. Approximately 25 percent of the workplace accidents that occur each year as the result of acceleration and impact involve objects that become projectiles.

Protection from Falling/Accelerating Objects

Objects that fall, are slung from a machine, or otherwise become projectiles pose a serious hazard to the heads, faces, feet, and eyes of workers. Consequently, protecting workers from projectiles requires the use of appropriate personal protective equipment and strict adherence to safety rules by all employees.

Head Protection

Approximately 120,000 people sustain head injuries on the job each year.[11] Falling objects are involved in many of these accidents. These injuries occur in spite of the fact that many of the victims were wearing hard hats. Such statistics have been the driving force behind the development of tougher, more durable hard hats.

According to Bross, "For nearly 70 years, the conventional hard hat—hard outershell and absorbing suspension system—has succeeded in helping prevent injuries at worksites across the nation and around the world."[12] Originally introduced in 1919, the hard hats first used for head protection in an industrial setting were inspired by the helmets worn by soldiers in World War I. Such early versions were made of varnished resin-impregnated canvas. As material technology evolved, hard hats were made of vulcanized fiber, then aluminum, and then fiberglass. Today's hard hats are typically made from the thermoplastic material polyethylene, using the injection-molding process.[13] Basic hard hat design has not changed radically since before World War II. They are designed to provide limited protection from impact primarily to the top of the head, and thereby reduce the amount of impact transmitted to the head, neck, and spine.[14]

The American National Standards Institute (ANSI) standard for hard hats is Z89–1986. OSHA subsequently adopted this standard as its hard hat standard (29 C.F.R. 1010.135). According to Bross,

> This standard calls for testing hard hats for impact attenuation and penetration resistance as well as electrical insulation. Specifically, hard hats are tested to withstand a 40-foot-pound impact, which is equivalent to a two-pound hammer falling about 20 feet. Hard hats are also designed to limit penetration of sharp objects that may hit the top of the hard-hat shell and to provide some lateral penetration protection.[15]

Hard hats can help reduce the risk associated with falling or projected objects, but only if they are worn. The use of hard hats in industrial settings that might have falling objects has been mandated by federal law since 1971.[16] In addition to making the use of hard hats mandatory when appropriate and supervising to ensure compliance, Feuerstein recommends the use of incentives.[17] According to Feuerstein,

> It would seem that the sweetest offer a head-injury prevention program makes is a work environment free of injuries from falling objects. But sometimes this ultimate reward is too abstract to excite employees. They need to be led into safety for its own sake by concrete incentives, such as intra-department competition, monetary rewards for good suggestions, points toward prizes, and peer recognition for the most improved behavior.[18]

Resources expended promoting the use of hard hats are resources wisely invested. "Work accidents resulting in head injuries cost employers and workers an estimated $2.5 billion per year in workers' compensation insurance, medical expenses and accident investigation as well as associated costs due to lost time on the job and substitute workers. That is an average cost of $22,500 for each worker who received a head injury."[19]

Eye and Face Protection

Eye and face protection typically consists of safety goggles or face shields. The ANSI standard for face and eye protective devices is Z87.1–1989. OSHA has also adopted this standard. It requires that nonprescription eye and face protective devices pass two impact tests: a high mass, low-speed test and a low mass, high-speed test.

The high-mass impact test determines the level of protection provided by face and eye protective devices from relatively heavy, pointed objects that are moving at low speeds. The high-speed impact test determines the level of protection provided from low-mass objects moving at high velocity.

Foot Protection

The OSHA regulations for foot protection are found in 29 C.F.R. 1910.132 and 126. Foot and toe injuries account for almost 20 percent of all disabling workplace injuries in the United States.[20] There are over 100,000 foot and toe injuries in the workplace each year.[21] According to Kelly, the major kinds of injuries to the foot and toes are from

- Falls/impact from sharp and/or heavy objects (this type accounts for 60 percent of all injuries)
- Compression when rolled over by or pressed between heavy objects
- Punctures through the sole of the foot
- Conductivity of electricity or heat
- Electrocution from contact with an energized, conducting material
- Slips on unstable walking surfaces
- Hot liquid or metal splashed into shoes or boots
- Temperature extremes[22]

The key to protecting workers' feet and toes is to match the protective measure with the hazard. This involves the following steps: (1) identify the various types of hazards present in the workplace, (2) identify the types of footwear available to counter the hazards, and (3) require that proper footwear be worn. Shoes selected should meet all applicable ANSI standards and have a corresponding ANSI rating. For example, "A typical ANSI rating is Z41PT83M1–75C–25. This rating means that the footwear meets the 1983 ANSI standard and the steel toe cap will withstand 75 foot pounds of impact and 2,500 pounds of compression."[23]

Modern safety boots are available that provide comprehensive foot and toe protection. The best safety boots will provide all of the following types of protection:

- *Steel toe* for impact protection
- *Rubber or vinyl* for chemical protection
- *Puncture-resistant soles* for protection against sharp objects
- *Slip-resistant soles* for protection against slippery surfaces
- *Electricity-resistant material* for protection from electric shock.

Employers are not required to provide footwear for employees, but they are required (29 C.F.R. 1910.132 and 136) to provide training on foot protection. The training must cover the following topics as a minimum:

- Conditions when protective footwear should be worn
- Type of footwear needed in a given situation
- Limitations of protective footwear
- Proper use of protective footwear

LIFTING HAZARDS

Back injuries that result from improper lifting are among the most common in an industrial setting. In fact, back injuries account for approximately $12 billion in workers' compensation costs annually. Putnam relates the following statistics concerning workplace back injuries:

- Lower back injuries account for 20 to 25 percent of all workers' compensation claims.
- Thirty-three to 40 percent of all worker's compensation costs are related to lower back injuries.

- Each year there are approximately 46,000 back injuries in the workplace.
- Back injuries cause 100 million lost workdays each year.
- Approximately 80 percent of the population will experience lower back pain at some point in their lives.[24]

Back injuries in the workplace are typically caused by improper lifting, reaching, sitting, and bending. Lifting hazards such as poor posture, ergonomic factors, and personal lifestyles also contribute to back problems. Consequently, a company's overall safety and health program should have a back safety/lifting component.

Back Safety/Lifting Program[25]

Prevention is critical in back safety. Consequently, safety and health professionals need to know how to establish back safety programs that overcome the hazards of lifting and other activities. Dr. Alex Kaliokin recommends the following six-step program:

1. *Display poster illustrations.* Posters that illustrate proper lifting, reaching, sitting, and bending techniques should be displayed strategically throughout the workplace. This is as important in offices as in the plant. Clerical and office personnel actually sustain a higher proportion of back injuries than employees in general. Sitting too long without standing, stretching, and walking can put as much pressure on the back as lifting.

2. *Preemployment screening.* Preemployment screening can identify people who already have back problems when they apply. This is important because more than 40 percent of back injuries occur in the first year of employment and the majority of these injuries are back related.

3. *Regular safety inspections.* Periodic inspections of the workplace can identify potential problem areas so that corrective action can be taken immediately. Occasionally bringing a workers' compensation consultant in to assist with an inspection can help identify hazards that company personnel might miss.

4. *Education and training.* Education and training designed to help employees understand how to lift, bend, reach, stand, walk, and sit safely can be the most effective preventive measure undertaken. Companies that provide back safety training report a significant decrease in back injuries.

5. *Use external services.* A variety of external health-care agencies can help companies extend their programs. Identify local health-care-providing agencies and organizations, what services they can provide, and a contact person in each. Maintaining a positive relationship with these external service contact people can increase the services available to employers.

6. *Map out the prevention program.* The first five steps should be written down and incorporated in the company's overall safety and health program. The written plan should be reviewed periodically and updated as needed.

In spite of a company's best efforts, back injuries will still occur. Consequently, modern safety and health managers should be familiar with the treatment and therapy that injured employees are likely to receive. According to Putnam, "Aggressive treatment for reconditioning addresses five goals: restoring function, reducing pain, minimizing deficits in strength, reducing lost time, and returning the body to pre-injury fitness levels."[26]

A concept that is gaining acceptance in bridging the gap between treatment/therapy and a safe return to work is known as work hardening.[27] Putnam describes work hardening as follows:

In specifically designed facilities known as "work centers" a broad range of work stations, exercise equipment, and aggressive protocols can focus on work-reconditioning. The objectives are:

- A return to maximum physical abilities;
- Improvement of general body fitness;
- Training to limit the possibility of re-injury;
- Work simulation that duplicates demands.[28]

The work centers referred to by Putnam replicate in as much detail as possible the injured employee's actual work environment. In addition to undergoing carefully controlled and monitored therapy in the work center, the employee is encouraged to use exercise equipment. Employees who undergo work center therapy should have already completed a program of acute physical therapy and pain management and they should be medically stable.[29]

Health and safety managers can help facilitate the fastest possible safe resumption of duties by injured employees by identifying local health-care providers that use the work-hardening approach. Such services and local providers of them should be made known to higher management so that the company can take advantage of them.

Proper Lifting Techniques

One of the most effective ways to prevent back injuries is to teach employees proper lifting techniques. Figure 7–1 summarizes lifting techniques that should be taught as part of an organization's safety program.

STANDING HAZARDS

Consider this statement by Roberta Carson, a certified professional ergonomist: "Prolonged standing or walking is common in industry and can be very painful. Low back pain, sore feet, varicose veins, swelling in the legs, general muscular fatigue, and other health problems have been associated with prolonged standing or walking."[30] Carson recommends the following precautions for minimizing standing hazards:[31]

Checklist of Lifting Techniques

Plan Ahead

✓ Determine if you can lift the load. Is it too heavy or too awkward?

✓ Decide if you need assistance.

✓ Check your route to see whether it has obstructions and slippery surfaces.

Lift with Your Legs, Not Your Back

✓ Bend at your knees, keeping your back straight.

✓ Position your feet close to the object.

✓ Center your body over the load.

✓ Lift straight up smoothly; don't jerk.

✓ Keep your torso straight; don't twist while lifting or after the load is lifted.

✓ Set the load down slowly and smoothly with a straight back and bent knees; don't let go until the object is on the floor.

Push, Don't Pull

✓ Pushing puts less strain on your back; don't pull objects.

✓ Use rollers under the object whenever possible.

Figure 7–1
Lifting techniques for preventing injuries.

Anti-Fatigue Mats

Anti-fatigue mats provide cushioning between the feet and hard working surfaces such as concrete floors. This cushioning effect can reduce muscle fatigue and lower back pain. However, too much cushioning can be just as bad as too little. Consequently, it is important to test mats on a trial basis before buying a large quantity. Mats that become slippery when wet should be avoided. In areas where chemicals are used, be sure to select mats that will hold up to chemicals.

Shoe Inserts

When anti-fatigue mats are not feasible because employees must move from area to area and, correspondingly, from surface to surface, shoe inserts may be the answer. Such inserts are worn inside the shoe and provide the same type of cushioning the mats provide. Shoe inserts can help reduce lower back, foot, and leg pain. It is important to ensure proper fit. If inserts make an employee's shoes too tight, they will do more harm than good. In such cases, employees may need to wear a slightly larger shoe size.

Foot Rails

Foot rails added to work stations can help relieve the hazards of prolonged standing. Foot rails allow employees to elevate one foot at a time four or five inches. The elevated foot rounds out the lower back, thereby relieving some of the pressure on the spinal column. Placement of a rail is important. It should not be placed in a position that inhibits movement or becomes a tripping hazard.

Workplace Design

A well-designed workstation can help relieve the hazards of prolonged standing. The key is to design workstations so that employees can move about while they work and can adjust the height of the workstation to match their physical needs.

Sit/Stand Chairs

Sit/stand chairs are higher-than-normal chairs that allow employees who typically stand while working to take quick mini-breaks and return to work without the hazards associated with getting out of lower chairs. They have the advantage of giving the employee's feet, legs, and back an occasional rest without introducing the hazards associated with lower chairs.

Proper Footwear

Proper footwear is critical for employees who stand for prolonged periods. Well-fitting, comfortable shoes that grip the worksurface and allow free movement of the toes are best.

Figure 7–2 can be used as a practical tool for identifying fall and lifting hazards.

APPLICATION SCENARIOS

1. Develop a comprehensive slip and fall prevention program that can be used by any small manufacturing firm.
2. Your company requires hard hats, but not the kind that have side protection. Write a rationale for requiring side protection.
3. Develop a comprehensive back/lifting safety program for a medium-sized manufacturing firm (800 employees).

Checklist for Identifying Fall and Lifting Hazards

Fall-Related Hazards

✓ Are foreign objects present on the walking surface or in walking paths?

✓ Are there design flaws in the walking surface?

✓ Are there raised or lowered sections of the walking surface that may trip a worker?

✓ Is good housekeeping being practiced?

✓ Is the walking surface made of or covered with a nonskid material?

✓ Are employees wearing nonskid footwear as appropriate?

✓ Are ladders strong enough to support the loads to which they are subjected?

✓ Are ladders free of cracks, loose rungs, and damaged connections?

✓ Are ladders free of heat damage and corrosion?

✓ Are ladders free of moisture that can cause them to conduct electricity?

✓ Are metal ladders free of burrs and sharp edges?

✓ Are fiberglass ladders free of blooming damage?

✓ Are employees wearing head and face protection as appropriate?

✓ Are employees wearing foot protection as appropriate?

Lifting Hazards

✓ Are posters that illustrate proper lifting techniques displayed strategically throughout the workplace?

✓ Are machines and other lifting aids available to assist employees in situations where loads to be lifted are too heavy or bulky?

✓ Are employees who are involved in lifting using personal protective devices?

Figure 7–2
A convenient tool for identifying fall and lifting hazards.

ENDNOTES

1. Kohr, R. L. "Slip Slidin' Away," *Safety & Health*, November 1989, Vol. 140, No. 5, p. 52.
2. Ibid.
3. Miller, B. C. "Falls: A Cast of Thousands Cost of Millions," *Safety & Health*, February 1988, Vol. 137, No. 2, p. 24.
4. Ibid.
5. Goldsmith, A. "Natural Walking, Unnatural Falls," *Safety & Health,* December 1988, Vol. 138, No. 6, p. 44.
6. Ibid.
7. Kohr, R. L. "Slip Slidin' Away," p. 52.
8. National Safety Council. "Ladder Safety Is No Accident," *Today's Supervisor*, June 1991, pp. 8–9.
9. Ibid., p. 8.
10. Ibid., p. 9.
11. Bross, M. S. "Advances Lead to Tougher, More Durable Hard Hat," *Occupational Health & Safety*, February 1991, Vol. 60, No. 2, p. 22.
12. Ibid.
13. Ibid.

14. Ibid., p. 23.

15. Ibid.

16. Feuerstein, P. "Head Protection Looks Up," *Safety & Health,* September 1991, Vol. 144, No. 3, p. 38.

17. Ibid., p. 39.

18. Ibid.

19. Bross, M. S. "Advances Lead to Tougher, More Durable Hard Hat," p. 23.

20. Dutton, C. "Make Foot Protection a Hit," *Safety & Health,* November 1988, Vol. 138, No. 5, p. 30.

21. Ibid.

22. Kelly, S. M. "Start Out with the Right Footwear," *Safety & Health,* February 1990, Vol. 141, No. 2, pp. 48–49.

23. Dutton, C. "Make Foot Protection a Hit," pp. 31–32.

24. Putnam, A. "How to Reduce the Cost of Back Injuries," *Safety & Health,* October 1988, Vol. 138, No. 4, pp. 48–49.

25. Kaliokin, A. "Six Steps Can Help Prevent Back Injuries and Reduce Compensation Costs," *Safety & Health,* October 1988, Vol. 138, No. 4, p. 50.

26. Putnam, A. "How to Reduce the Cost of Back Injuries," p. 50.

27. Ibid.

28. Ibid.

29. Ibid.

30. Carson, R. "Stand By Your Job," *Occupational Health & Safety,* April 1994, p. 38.

31. Ibid., pp. 40–42.

CHAPTER EIGHT

Temperature and Pressure Hazards

Part of providing a safe and healthy workplace is appropriately controlling the temperature, humidity, air distribution and pressure in work areas. A work environment in which the temperature and pressure are not properly controlled can be uncomfortable. Extremes of heat, cold, and pressure can be more than uncomfortable—they can be dangerous. Heat stress, cold stress, burns, anoxia, and hypoxia are major concerns of modern safety and health professionals. This chapter provides the information that professionals need to know to overcome the hazards associated with temperature and pressure hazards.

THERMAL COMFORT[1]

Thermal comfort in the workplace is a function of a number of different factors. Temperature, humidity, air distribution, personal preference, and acclimatization are all determinants of comfort in the workplace. However, determining optimum conditions is not a simple process.

To fully understand the hazards posed by temperature extremes, safety and health professionals must be familiar with several basic concepts related to thermal energy. The most important of these are summarized here:

- Conduction is the transfer of heat between two bodies that are touching, or from one location to another within a body. For example, if an employee touches a workpiece that has just been welded and is still hot, heat will be conducted from the workpiece to the hand. Of course, the result of this heat transfer is a burn.

- Convection is the transfer of heat from one location to another by way of a moving medium (a gas or a liquid). Convection ovens use this principle to transfer heat from an electrode by way of gases in the air to whatever is being baked.

- Metabolic heat is produced within a body as a result of activity that burns energy. All humans produce metabolic heat. This is why a room that is comfortable when occupied by just a few people may become uncomfortable when it is crowded. Unless the thermostat is lowered to compensate, the metabolic heat of a crowd will cause the temperature of a room to rise to an uncomfortable level.

- Environmental heat is produced by external sources. Gas or electric heating systems produce environmental heat as do sources of electricity and a number of industrial processes.

- Radiant heat is the result of electromagnetic non-ionizing energy that is transmitted through space without the movement of matter within that space.

THE BODY'S RESPONSE TO HEAT

The human body is equipped to maintain an appropriate balance between the metabolic heat that it produces and the environmental heat to which it is exposed. Sweating and the subsequent evaporation of the sweat are the body's way of trying to maintain an acceptable temperature balance.

As long as heat gained from radiation, convection, and metabolic processes does not exceed that lost through the evaporation induced by sweating, the body experiences no stress or hazard. However, when heat gain from any source or sources is more than the body can compensate for by sweating, the result is heat stress.

Heat stress can manifest itself in a number of ways depending on the level of stress. The most common types of heat stress are heat stroke, heat exhaustion, heat cramps, heat rash, transient heat fatigue, and chronic heat fatigue. These various types of heat stress can cause a number of undesirable bodily reactions including prickly heat, inadequate venous return to the heart, inadequate blood flow to vital body parts, circulatory shock, cramps, thirst, and fatigue. Different types of heat stress and how to prevent them are covered in the next section.

HEAT STRESS AND ITS PREVENTION

Heat stress is a major concern of modern safety and health professionals. According to North, "Ads on television and in magazines glorify heat. They say: 'Visit tropic lands, bask in the hot sun while wearing the newest tanning lotion, enjoy a home sauna and live in the sun belt.' But in spite of all the messages to the contrary, heat can be harmful—even deadly, under extreme conditions."[2] This section covers various types of heat stress, how it can be prevented, and actions to take when heat stress occurs in spite of preventive measures.

Heat Stroke: Cause, Symptoms, Treatment, and Prevention

Heat stroke is a type of heat stress that occurs as a result of a rapid rise in the body's core temperature. Heat stroke is very dangerous and should be dealt with immediately because it can be fatal. One can recognize heat stroke by observing the symptoms shown in Figure 8–1: (1) hot, dry, mottled skin, (2) confusion and/or convulsions, and (3) loss

Figure 8–1
Observable symptoms of heat stroke.

> **Checklist of Heat Stroke Symptoms**
>
> ✓ Skin is hot, dry, and typically red and mottled
>
> ✓ Loss of consciousness
>
> ✓ Confusion and/or convulsions

of consciousness. In addition to these observable symptoms, a victim of heat stroke will have a rectal temperature of 104.5°F or higher that will typically continue to climb.

Several factors can make an individual susceptible to heat stroke. These factors include the following: (1) obesity, (2) poor physical condition, (3) alcohol intake, (4) cardiovascular disease, and (5) prolonged exertion in a hot environment. This last factor can cause heat stroke even in a healthy individual. An employee who has one or more of the first four characteristics is even more susceptible to heat stroke.

In cases of heat stroke, the body's ability to sweat becomes partially impaired or actually breaks down altogether. This sets in motion a situation in which the temperature begins to increase uncontrollably. If this situation is not reversed quickly, heat stroke can be fatal.

If an employee becomes a heat stroke victim, action must be taken immediately to reduce his or her body core temperature. Do not wait until medical help arrives to begin. The victim should be immersed in chilled water if facilities are available. If not, wrap the victim in a wet, thin sheet and fan continuously, adding water periodically to keep the sheet wet.

Prevention strategies include the following: (1) medical screening as part of the employment process to identify applicants who have one or more susceptibility characteristics; (2) gradual acclimatization to hot working conditions spread over at least a full week; (3) rotation of workers out of the hot environment at specified intervals during the work day; (4) use of personal protective clothing that is cooled; and (5) monitoring employees carefully and continually.

An example of using special clothing can be found at the nuclear power plant Three Mile Island in Unit 2. According to Hildebrand, during the summer months, the temperature in Unit 2 can exceed 90°F, making it difficult for employees to work for sustained periods. To counter this problem, Unit 2's safety and health team implemented a program that involves the use of special self-cooling clothing of the following two types: (1) an ice vest that consists of a light fitting vest with 60 small pockets of ice with a total ice capacity of about eight pounds; and (2) a vinyl one-piece coverall with a built-in air distribution system that directs cool air against the body and exhausts warm air. The ice vest is used when mobility is important and the length of exposure is relatively short. The body suit is used when a longer exposure time is necessary.[3]

Heat Exhaustion: Cause, Symptoms, Treatment, and Prevention

Heat exhaustion is a type of heat stress that occurs as a result of water and/or salt depletion. Figure 8–2 summarizes the observable symptoms of heat exhaustion. In addition to these symptoms, a victim of heat exhaustion may have a normal or even lower-than-normal oral temperature, but will typically have a higher-than-usual rectal temperature (i.e., 99.5°F to 101.3°F or 37.5°C to 38.5°C).

Heat exhaustion can be brought on by prolonged exertion in a hot environment and a failure to replace the water and/or salt lost through sweating. It can be compounded by a failure to acclimate employees gradually to working in a hot environment.

Figure 8–2
Observable symptoms of heat exhaustion.

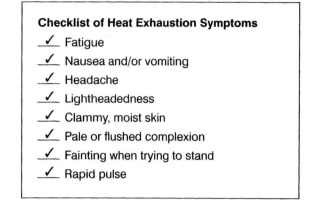

Checklist of Heat Exhaustion Symptoms
✓ Fatigue
✓ Nausea and/or vomiting
✓ Headache
✓ Lightheadedness
✓ Clammy, moist skin
✓ Pale or flushed complexion
✓ Fainting when trying to stand
✓ Rapid pulse

In cases of heat exhaustion, the body becomes dehydrated. This, in turn, decreases the volume of blood circulating. The various body parts must then compete for a smaller volume of blood. This causes circulatory strain, which manifests itself in the types of symptoms summarized in Figure 8–2.

A victim of heat exhaustion should be moved to a cool, but not cold, environment and allowed to rest lying down. Fluids should be taken slowly but steadily by mouth until the urine volume indicates that the body's fluid level is once again in balance.

Prevention of heat exhaustion should be handled on the job in the same way as it is at athletic events. Professional baseball and football teams have spring training and a preseason to help get players acclimated to working conditions. Gradual acclimatization over at least a week is also important for employees who will work in the heat. During games, professional athletes replace the fluids that they lose by drinking one of the drinks commercially produced for this purpose. Such drinks are better than water because they contain appropriate amounts of salt, electrolytes, and other important elements lost during sweating. Employees working in the heat should have such fluids readily available and drink them frequently.

Electrolyte imbalance is a problem with heat exhaustion and heat cramps. When people sweat in response to exertion and environmental heat, they lose more than just water. They also lose salt and electrolytes. Electrolytes are minerals that are needed for the body to maintain its proper metabolism and for cells to produce energy. Loss of electrolytes causes these functions to break down. For this reason, it is important to use commercially produced drinks that contain water, salt, sugar, potassium, or electrolytes to replace those lost through sweating.

Heat Cramps: Cause, Symptoms, Treatment, and Prevention

Heat cramps are a type of heat stress that occur as a result of salt and potassium depletion. Observable symptoms are primarily muscle spasms that are typically felt in the arms, legs, and abdomen. Heat cramps are caused by salt and potassium depletion from profuse sweating as a result of working in a hot environment. Drinking water without also replacing salt exacerbates the problem.

During heat cramps, salt is lost, water that is taken in dilutes the body's electrolytes, and excess water enters the muscles causing cramping. When heat cramps occur, the appropriate response is to replenish the body's salt and potassium supply orally. This can be done with commercially produced fluids that contain carefully measured amounts of salts, potassium, electrolytes, and other elements.

To prevent heat cramps, acclimate workers to the hot environment gradually over a period of at least a week. Then, ensure that fluid replacement is accomplished with a product that contains the appropriate amount of salt, potassium, and electrolytes.

Heat Rash: Cause, Symptoms, Treatment, and Prevention

Heat rash is a type of heat stress that manifests itself as small, raised bumps or blisters that cover a portion of the body and give off a prickly sensation that can cause discomfort. It is caused by prolonged exposure to hot and humid conditions in which the body is continuously covered with sweat that does not evaporate because of the high humidity. The sweat gland ducts become clogged with retained sweat that does not evaporate. The sweat backs up in the system and causes minor inflammation.

Heat rash is simply and easily treated by removing the victim to a cooler, less humid environment, cleaning the affected area, and changing wet clothes for dry. Special lotions are available to speed the healing process. Heat rash can be prevented by resting in a cool, nonhumid environment and periodically changing into dry clothing.

Heat Fatigue: Cause, Symptoms, Treatment, and Prevention

Transient heat fatigue is a form of heat stress that manifests itself in temporary sluggishness, lethargy, and impaired performance (mental and/or physical). Employees who are not acclimated to working in a hot environment are especially susceptible to transient heat fatigue. The degree and frequency of transient heat fatigue is also a function of physical conditioning.

Well-conditioned employees who are properly acclimated will suffer this form of heat stress less frequently and less severely than poorly conditioned employees will. Consequently, preventing transient heat fatigue involves physical conditioning and acclimatization.

Chronic heat fatigue is similar to transient heat fatigue except that it does not abate after an appropriate rest period. Employees who experience chronic heat fatigue should be moved into jobs that do not involve working in a hot environment. Prolonged chronic heat fatigue, if not relieved, can cause both physiological and psychological stress. The psychological stress can manifest itself in substance abuse and other psychosocially unacceptable behavior.

BURNS AND THEIR EFFECTS

One of the most common hazards associated with heat in the workplace is the burn. Burns can be especially dangerous because they disrupt the normal functioning of the skin, which is the body's largest organ and the most important in terms of protecting other organs. It is necessary first to understand the composition of, and purpose served by, the skin to understand the hazards that burns can represent.

Human Skin

Human skin is the tough, continuous outer covering of the body. It consists of the following two main layers: (1) the outer layer, which is known as the epidermis; and (2) the inner layer, which is known as the dermis, cutis, or corium. The dermis is connected to the underlying subcutaneous tissue.

The skin serves several important purposes including the following: protection of body tissue, sensation, secretion, excretion, and respiration. Protection from fluid loss, water penetration, ultraviolet radiation, and infestation by microorganisms is a major function of the skin. The sensory functions of touching, sensing cold, feeling pain, and sensing heat involve the skin.

The skin helps regulate body heat through the sweating process. It excretes sweat that takes with it electrolytes and certain toxins. This helps keep the body's fluid level in balance. By giving off minute amounts of carbon dioxide and absorbing small amounts of oxygen, the skin also aids slightly in respiration.

What makes burns particularly dangerous is that they can disrupt any or all of these functions depending on their severity. The deeper the penetration, the more severe the burn.

Severity of Burns

The severity of a burn depends on several factors. The most important of these is the depth to which the burn penetrates. Other determining factors include location of the burn, age of the victim, and amount of burned area.

The most widely used method of classifying burns is by degree (i.e., first-, second-, or third-degree burns). Modern safety and health professionals should be familiar with these classifications and what they mean.

First-degree burns are minor and result only in a mild inflammation of the skin, known as *erythema*. Sunburn is a common form of first-degree burn. It is easily recognizable as a redness of the skin that makes the skin sensitive and moderately painful to the touch.

Second-degree burns are easily recognizable from the blisters that form on the skin. If a second-degree burn is superficial, the skin will heal with little or no scarring. A deeper second-degree burn will form a thin layer of coagulated, dead cells that feels leathery to the touch. A temperature of approximately 210°F can cause a second-degree burn in as little as 15 seconds of contact.

Third-degree burns are very dangerous and can be fatal depending on the amount of body surface affected. A third-degree burn penetrates through both the epidermis and the dermis. A deep third-degree burn will penetrate body tissue. Third-degree burns can be caused by both moist and dry hazards. Moist hazards include steam and hot liquids; these cause burns that appear white. Dry hazards include fire and hot objects or surfaces; these cause burns that appear black and charred.

In addition to the depth of penetration of a burn, the amount of surface area covered is also a critical concern. This amount is expressed as a percentage of body surface area, or BSA. Figure 8–3 shows how the percentage of BSA can be estimated. Burns covering over 75 percent of BSA are usually fatal.

Using the first-, second-, and third-degree burn classifications in conjunction with BSA percentages, burns can be classified further as minor, moderate, or critical. According to Blocker, these classifications can be summarized as described in the following paragraphs.[4]

Minor Burns

All first-degree burns are considered minor. Second-degree burns covering less than 15 percent of the body are considered minor. Third-degree burns can be considered minor provided they cover only 2 percent or less of BSA.

Figure 8–3
Estimating percentage of body surface area (BSA) burned.

Checklist for Estimating BSA

✓ Right arm ....9% of BSA

✓ Left arm....................................9% of BSA

✓ Head/neck ....9% of BSA

✓ Right leg...................................18% of BSA

✓ Left leg ....18% of BSA

✓ Back...18% of BSA

✓ Chest/stomach ....18% of BSA

✓ Perineum...................................1% of BSA

Moderate Burns

Second-degree burns that penetrate the epidermis and cover 15 percent or more of BSA are considered moderate. Second-degree burns that penetrate the dermis and cover from 15 to 30 percent of BSA are considered moderate. Third-degree burns can be considered moderate provided they cover less than 10 percent of BSA and are not on the hands, face, or feet.

Critical Burns

Second-degree burns covering more than 30 percent of BSA or third-degree burns covering over 10 percent of BSA are considered critical. Even small-area third-degree burns to the hands, face, or feet are considered critical because of the greater potential for infection to these areas by their nature. In addition, burns that are complicated by other injuries (fractures, soft tissue damage, and so on) are considered critical.

CHEMICAL BURNS

Chemicals are widely used in modern industry even by companies that do not produce them as part of their product base. Many of the chemicals produced, handled, stored, transported and/or otherwise used in industry can cause burns similar to those caused by heat (i.e., first-, second-, and third-degree burns). The hazards of chemical burns are very similar to those of thermal burns.

Chemical burns, like thermal burns, destroy body tissue; the extent of destruction depends on the severity of the burn. However, chemical burns continue to destroy body tissue until the chemicals are washed away completely.

The severity of the burn produced by a given chemical depends on the following factors:

- Corrosive capability of the chemical
- Concentration of the chemical
- Temperature of the chemical or the solution in which it is dissolved
- Duration of contact with the chemical[5]

Checklist of Harmful Effects of Chemicals	
Chemical	**Potential Harmful Effect**
✓ Acetic Acid	Tissue damage
✓ Liquid bromide	Corrosive effect on the respiratory system and tissue damage
✓ Formaldehyde	Tissue hardening
✓ Lime	Dermatitis and eye burns
✓ Methylbromide	Blisters
✓ Nitric/sulfuric acid mixture	Severe burns and tissue damage
✓ Oxalic acid	Ulceration and tissue damage
✓ White phosphorus	Ignites in air causing thermal burns
✓ Silver nitrate	Corrosive/caustic effect on the skin
✓ Sodium (metal)	Ignites with moisture causing thermal burns
✓ Trichloracetic acid	Tissue damage

Figure 8–4
Harmful effects of selected widely used chemicals.

Effects of Chemical Burns

Different chemicals have different effects on the human body. The harmful effects of selected widely used chemicals are summarized in Figure 8–4.[6] These are only a few of the many chemicals widely used in industry today. All serve an important purpose; however, all carry the potential for serious injury.

The primary hazardous effects of chemical burns are infection, loss of body fluids, and shock, and are summarized in the following paragraphs.[7]

Infection

The risk of infection is high with chemical burns—as is it with heat-induced burns—because the body's primary defense against infection-causing microorganisms (the skin) is penetrated. This is why it is so important to keep burns clean. Infection in a burn wound can cause *septicemia* (blood poisoning).

Fluid Loss

Body fluid loss in second- and third-degree burns can be serious. With second-degree burns, the blisters that form on the skin often fill with fluid that seeps out of damaged tissue under the blister. With third-degree burns, fluids are lost internally and, as a result, can cause the same complications as a hemorrhage. If these fluids are not replaced properly, the burns can be fatal.

Shock

Shock is a depression of the nervous system. It can be caused by physical and/or psychological trauma. In cases of serious burns, shock may be caused by the intense pain that can occur when skin is burned away, leaving sensitive nerve endings exposed. Shock from burns can come in the following two forms: (1) primary shock, which is the first stage and results from physical pain and/or psychological trauma; and (2) secondary shock, which comes later and is caused by a loss of fluids and plasma proteins as a result of the burns.

First Aid for Chemical Burns

There is a definite course of action that should be taken when chemical burns occur, and the need for immediacy cannot be overemphasized. According to the National Safety Council, the proper response in cases of chemical burns "is to wash off the chemical by flooding the burned areas with copious amounts of water as quickly as possible. This is the only method for limiting the severity of the burn, and the loss of even a few seconds can be vital."[8]

In the case of chemical burns to the eyes, the continuous flooding should continue for at least 15 minutes. The eyelids should be held open to ensure that chemicals are not trapped under them.

Another consideration when an employee comes in contact with a caustic chemical is his or her clothing. If chemicals have saturated the employee's clothes, they must be removed quickly. The best approach is to remove the clothes while flooding the body or the affected area. If necessary for quick removal, clothing should be ripped or cut off.

The critical need to apply water immediately in cases of chemical burns means that water must be readily available. Health and safety professionals should ensure that special eye wash and shower facilities are available wherever employees handle chemicals.

OVERVIEW OF COLD HAZARDS

Temperature hazards are generally thought of as extremes of heat. This is natural because most workplace temperature hazards do relate to heat. However, temperature extremes at the other end of the spectrum—cold—can also be hazardous. Employees who work outdoors in colder climates and employees who work indoors in such jobs as meatpacking are subjected to cold hazards.

The major injuries associated with extremes of cold can be classified as being either generalized or localized. A generalized injury from extremes of cold is hypothermia. Localized injuries include frostbite, frostnip, and trenchfoot. According to Alpaugh, "The main factors contributing to cold injury are exposure to humidity and high winds, contact with wetness or metal, inadequate clothing, age, and general health. Physical conditions that worsen the effects of cold include allergies, vascular disease, excessive smoking and drinking, and specific drugs and medicine."[9]

The wind-chill factor increases the level of hazard posed by extremes of cold. Health and safety professionals need to understand this concept and how to make it part of their deliberations when developing strategies to prevent cold stress injuries.

Wind-Chill Factor

The human body is able to sense cold. What the body actually senses when it is cold is a combination of temperature and wind velocity. Wind or air movement causes the body to sense coldness beyond what the thermometer actually registers as the temperature. This phenomenon is known as the wind-chill factor. It is the cooling effect produced by a combination of temperature, wind velocity, and/or air movement.[10]

When the temperature and the wind-chill factor are considered together, the result is an equivalent temperature that is lower than the thermometer reading. For example, an actual temperature reading of 10°F coupled with a 15-mph wind results in an equivalent temperature of –18°F. Wind increases the hazard of cold stress noticeably.

Frostbite, Frostnip, and Trenchfoot

The less severe disorders that can result from cold stress are frostbite, frostnip, and trenchfoot.

Frostbite is similar to burns in that it has three degrees. With first-degree frostbite, there is freezing but no blistering or peeling. With second-degree frostbite, there is freezing accompanied by blistering and peeling. With third-degree frostbite, there is freezing accompanied by death of skin and/or tissue. The first sign of frostbite is typically a sensation of cold and numbness. These symptoms may be accompanied by tingling, stinging, aching, or cramps. Frostbite of the outer layer of skin results in a whitish, waxy look. Deep frostbite results in tissue that is cold, pale, and solid.[11]

Frostnip is less severe than frostbite. It causes the skin to turn white and typically occurs on the face and other exposed parts of the body. There is no tissue damage with frostnip. However, if the exposed area is not either covered or removed from exposure to the cold, frostnip can become frostbite.

Trenchfoot is a condition that manifests itself as a tingling, itching, swelling, and pain. If these symptoms are not treated, this condition can lead to more serious injury including blistering, death of tissue, and ulceration. Trenchfoot is caused by continuous exposure of the feet simultaneously to a cold, but not freezing, environment and moisture.[12]

Hypothermia

Hypothermia is the condition that results when the body's core temperature drops to dangerously low levels. If this condition is not reversed, the patient literally freezes to death. Figure 8–5 summarizes the various symptoms of hypothermia that can be observed. The number of symptoms and their severity depend on how low the body's core temperature drops.

Figure 8–5
Observable symptoms of
hypothermia.

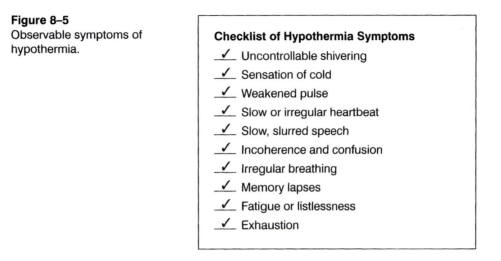

Checklist of Hypothermia Symptoms
✓ Uncontrollable shivering
✓ Sensation of cold
✓ Weakened pulse
✓ Slow or irregular heartbeat
✓ Slow, slurred speech
✓ Incoherence and confusion
✓ Irregular breathing
✓ Memory lapses
✓ Fatigue or listlessness
✓ Exhaustion

A person's susceptibility to hypothermia is increased by sedative drugs and alcohol. Sedatives interfere with the transmission of impulses from the nerve endings in the skin to the brain. This can cause a person to miss natural signals that he or she is in danger. Alcohol dilates blood vessels near the skin's surface. This, in turn, increases the amount and rate of heat loss, which results in an even lower body temperature.[13]

PREVENTING COLD STRESS

There are numerous strategies for preventing cold stress. Some of them are simple, common-sense strategies that employees should learn and practice. These include wearing appropriate protective clothing, limiting the duration of exposure to cold, replacing fluids (this is just as important in cold environments as it is in hot environments), eating a proper diet to ensure that the body is able to generate metabolic heat, and keeping the feet and all other extremities dry.

Cold Stress Prevention Program

Modern safety and health professionals in settings where cold stress is an issue should establish a cold stress prevention component as part of their overall safety and health program. The National Safety Council recommends that such a program contain the following elements:

1. *Medical supervision and screening.* Medical screening involves identifying individuals who are particularly susceptible to cold stress (i.e., applicants who are in poor physical condition, overweight, and/or have cardiovascular problems). Medical supervision involves medical checkups.

2. *Orientation and training.* Employees should learn about the hazards associated with extremes of cold and how to protect themselves and fellow workers, including the use of proper clothing, appropriate work scheduling, proper work practices, and first aid procedures.

3. *Work practices.* Employees should understand and use proper work practices including regularly scheduled fluid replacement and periodic rest breaks in a warm environment.

4. *Engineering and administrative controls.* Responsible company officials should reduce hazard levels as much as possible through the application of appropriate

engineering and administrative controls. Figure 8–6 summarizes engineering and administrative controls that can be used to decrease the potential for cold stress.[14]

PRESSURE HAZARDS DEFINED

Pressure is defined in physics as the force exerted against an opposing fluid or thrust distributed over a surface. This may be expressed in force or weight per unit of area, such as psi (pounds per square inch). A hazard is a condition with the potential of causing injury to personnel, damage to equipment or structures, loss of material, or lessening of the ability to perform a prescribed function. Thus, a pressure hazard is a hazard caused by a dangerous condition involving pressure. Critical injury and damage can occur with relatively little pressure. OSHA defines high-pressure cylinders as those designated with a service pressure of 900 psi or greater.

We perceive pressure in relation to the earth's atmosphere. Approximately 21 percent of the atmosphere is oxygen, with most of the other 79 percent being nitrogen. In addition to oxygen and nitrogen, the atmosphere contains trace amounts of several inert gases: argon, neon, krypton, xenon, and helium.

At sea level, the earth's atmosphere averages 1,013 ⟺ 10 N/m^2, or 1.013 millibars, or 760 mm Hg (29.92 inches), or 14.7 psi, depending on the measuring scale used.[15] The international system of measurement utilizes newtons per square meter (N/m^2). However, in human physiology studies, the typical unit is millimeters of mercury (mm Hg).

Atmospheric pressure is usually measured using a barometer. As the altitude above sea level increases, atmospheric pressure decreases in a nonlinear fashion. For example, at 18,000 feet above sea level, the barometric pressure is equal to 390 mm Hg. Half of this pressure, around 195 mm Hg, can be found at 23,000 feet above sea level.

Boyle's law states that the product of a given pressure and volume is constant with a constant temperature:

$$P_1 V_1 = P_2 V_2 \text{, when } T \text{ is constant}$$

Air moves in and out of the lungs because of a pressure gradient or difference in pressure. When atmospheric pressure is greater than pressure within the lungs, air flows

Figure 8-6
Checklist of engineering and administrative controls for preventing cold stress.

Checklist of Strategies for Preventing Cold Stress

✓ Use central or spot heating, warm air jets, contact warm plates, or radiant heaters to provide warmth.

✓ Shield the work area from wind.

✓ Cover the handles of metal tools with insulating material.

✓ Remove metal chairs or cover them with insulating material.

✓ Provide heated tents or shelter and require workers to use them periodically for warming breaks.

✓ Provide warmed drinks for fluid replacement (nonalcoholic, caffeine-free, properly balanced).

✓ Require an acclimatization period of all new workers.

✓ Use the buddy system and rotate the employees frequently.

✓ Design the job to minimize sitting or standing still.

✓ Design the job so that as many tasks as possible can be performed in a warm environment.

down this pressure gradient from the outside into the lungs. This is called inspiration, inhalation, or breathing in, and occurs with greater lung volume than at rest. When pressure in the lungs is greater than atmospheric pressure, air moves down a pressure gradient outward from the lungs to the outside. Expiration occurs when air leaves the lungs and the lung volume is less than the relaxed volume, increasing pressure within the lungs.

Gas exchange occurs between air in the lung alveoli and gas in solution in blood. The pressure gradients causing this gas exchange are called partial pressures. Dalton's law of partial pressures states that, in a mixture of theoretically ideal gases, the pressure exerted by the mixture is the sum of the pressures exerted by each component gas of the mixture:

$$P_A = P_O + P_N + P_{else}$$

Air entering the lungs immediately becomes saturated with water vapor. Water vapor, although it is a gas, does not conform to Dalton's law. The partial pressure of water vapor in a mixture of gases is not dependent on its fractional concentration in that mixture. Water vapor partial pressure, instead, is dependent on its temperature. From this exception to Dalton's law comes the fact that at the normal body temperature of 37YC, water vapor maintains a partial pressure of 47 mm Hg as long as that temperature is maintained. With this brief explanation of how pressure is involved in human breathing, we now focus on the various sources of pressure hazards.

SOURCES OF PRESSURE HAZARDS

There are many sources of pressure hazards—some natural, most created by humans. Since the human body is comprised of approximately 85 percent liquid, which is virtually incompressible, increasing pressure does not create problems by itself. Problems can result from air being trapped or expanded within body cavities.

When sinus passages are blocked so that air cannot pass easily from the sinuses to the nose, expansion of the air in these sinuses can lead to problems. The same complications can occur with air trapped in the middle ear's Eustachian tube. As Boyle's law states, gas volume increases as pressure decreases. Expansion of the air in blocked sinus passages or the middle ear occurs with a rapid increase in altitude or rapid ascent underwater. This can cause pain and, if not eventually relieved, disease. Under extreme circumstances of rapid ascent from underwater diving or high-altitude decompression, lungs can rupture.

Nitrogen absorption into the body tissues can become excessive during underwater diving and breathing of nitrogen-enriched air. Nitrogen permeation of tissues occurs in proportion to the partial pressure of nitrogen taken in. If the nitrogen is permeating tissues faster than the person can breathe it out, bubbles of gas may form in the tissues.

Decompression sickness can result from the decompression that accompanies a rapid rise from sea level to at least 18,000 feet or a rapid ascent from around 132 feet to 66 feet underwater. Several factors influence the onset of decompression sickness (see Figure 8–7).

A reduction in partial pressure can result from reduced available oxygen and cause a problem in breathing known as hypoxia. Too much oxygen or oxygen breathed under pressure that is too high is called hyperoxia. Another partial pressure hazard, nitrogen narcosis, results from a higher-than-normal level of nitrogen pressure.

When breathed under pressure, nitrogen causes a reduction of cerebral and neural activity. Breathing nitrogen at great depths underwater can cause a feeling of euphoria and loss of reality. At depths greater than 100 feet (30 meters), nitrogen narcosis can occur even when breathing normal air. The effects may become pathogenic at depths greater than 200 feet, with motor skills threatened at depths greater than 300 feet. Cognitive processes deteriorate quickly after reaching a depth of 325 feet.

Checklist of Factors That Influence the Onset of Decompression Sickness

✓ A *history* of previous decompression sickness increases the probability of another attack.

✓ *Age* is a component. Being over 30 increases the chances of an attack.

✓ *Physical fitness* plays a role. People in better condition have a reduced chance of the sickness. Previously broken bones and joint injuries are often the sites of pain.

✓ *Exercise* during the exposure to decompression increases the likelihood and brings on an earlier onset of symptoms.

✓ *Low temperature* increases the probability of the sickness.

✓ *Speed of decompression* also influences the sickness. A rapid rate of decompression increases the possibility and severity of symptoms.

✓ *Length of exposure* of the person to the pressure is proportionately related to the intensity of symptoms. The longer the exposure, the greater the chances of decompression sickness.

Figure 8-7
Decompression sickness; contributing factors.

BOILERS AND PRESSURE HAZARDS

A boiler is a closed vessel in which water is heated to form steam, hot water, or high-temperature water under pressure.[16] Potential safety hazards associated with boilers and other pressurized vessels include the following (see Figure 8–8).[17]

Through years of experience, a great deal has been learned about how to prevent accidents associated with boilers. The Traveler's Insurance Company recommends the following daily, weekly, monthly, and yearly accident prevention measures:

1. *Daily check.* Check the water to make sure that it is at the proper level. Vent the furnace thoroughly before starting the fire. Warm up the boiler using a small fire. When the boiler is operating, check it frequently.

2. *Weekly check.* At least once every week, test the low-water automatic shutdown control and record the results of the test on a tag that is clearly visible.

3. *Monthly check.* At least once every month, test the safety valve and record the results of the test on a tag that is clearly visible.

4. *Yearly check.* The low-level automatic shutdown control mechanism should be either replaced or completely overhauled and rebuilt. Arrange to have the vendor or a

Figure 8-8
Most common boiler hazards.

Checklist of Hazards Associated with Boilers

✓ Poor or insufficient training of operators

✓ Human error

✓ Mechanical breakdown/failure

✓ Failure or blockage of control and/or safety devices

✓ Insufficient or improper inspections

✓ Improper application of equipment

✓ Insufficient preventive maintenance

---

third-party expert test all combustion safeguards, including fuel pressure switches, limit switches, motor starter interlocks, and shutoff valves.[18]

HIGH-TEMPERATURE WATER HAZARDS[19]

High-temperature water (HTW) is exactly what its name implies—water that has been heated to a very high temperature, but not high enough to produce steam. In some cases, HTW can be used as an economical substitute for steam (e.g., in industrial heating systems). It has the added advantage of releasing less energy (pressure) than steam does.

In spite of this, there are hazards associated with HTW. Human contact with HTW can result in extremely serious burns and even death. The two most prominent sources of hazards associated with HTW are operator error and improper design. Proper training and careful supervision are the best guards against operator error.

Design of HTW systems is a highly specialized process that should be undertaken only by experienced engineers. Mechanical forces such as water hammer, thermal expansion, thermal shock, or faulty materials cause system failures more often than do thermodynamic forces. Therefore, it is important to allow for such causes when designing an HTW system.

The best designs are simple and operator-friendly. Designing too many automatic controls into an HTW system can create more problems than it solves by turning operators into mere attendants who are unable to respond properly to emergencies.

HAZARDS OF UNFIRED PRESSURE VESSELS[20]

Not all pressure vessels are fired. Unfired pressure vessels include compressed air tanks, steam-jacketed kettles, digesters, and vulcanizers, as well as others that can create heat internally by various means rather than by external fire. The various means of creating internal heat include (1) chemical action within the vessel, and (2) application of some heating medium (electricity, steam, hot oil, and so on) to the contents of the vessel. The potential hazards associated with unfired pressure vessels include hazardous interaction between the material of the vessel and the materials that will be processed in it; inability of the filled vessel to carry the weight of its contents and the corresponding internal pressure; inability of the vessel to withstand the pressure introduced into it plus pressure caused by chemical reactions that occur during processing; and inability of the vessel to withstand any vacuum that might be created accidentally or intentionally.

The most effective preventive measure for overcoming these potential hazards is proper design. Specifications for the design and construction of unfired pressure vessels include requirements in the following areas: working pressure range, working temperature range, type of materials to be processed, stress relief, welding/joining measures, and radiography. Designs that meet the specifications set forth for unfired pressure vessels in such codes as the ASME Code (Section VIII) will overcome most predictable hazards.

Beyond proper design, the same types of precautions taken when operating fired pressure vessels can be used when operating unfired vessels. These include continual inspection, proper housekeeping, periodic testing, visual observation (for detecting cracks), and the use of appropriate safety devices.

HAZARDS OF HIGH-PRESSURE SYSTEMS[21]

The hazards most commonly associated with high-pressure systems are leaks, pulsation, vibration, release of high-pressure gases, and whiplash from broken high-pressure pipe, tubing, or hose. Strategies for reducing these hazards include limiting vibration through the use of vibration dampening (use of anchored pipe supports); decreasing the potential for leaks by limiting the number of joints in the system; using pressure gauges; placing shields or barricades around the system; using remote control and monitoring; and restricting access.

PRESSURE DANGERS TO HUMANS

The term *anoxia* refers to the rare case of no oxygen being available. Hypoxia, a condition that occurs when the available oxygen is reduced, can occur while ascending to a high altitude or when oxygen in air has been replaced with another gas, which may happen in some industrial situations.

Altitude sickness is a form of hypoxia associated with high altitudes. Ascent to an altitude of 10,000 feet above sea level can result in a feeling of malaise, shortness of breath, and fatigue. A person ascending to 14,000 to 15,000 feet may experience euphoria, along with a reduction in powers of reason, judgment, and memory. Altitude sickness includes a loss of useful consciousness at 20,000 to 25,000 feet.[22] After approximately five minutes at this altitude, a person may lose consciousness. The loss of consciousness comes at approximately one minute or less at 30,000 feet. Over 38,000 feet, most people lose consciousness within 30 seconds and may fall into a coma and possibly die.

Hyperoxia, or an increased concentration of oxygen in air, is not a common situation. Hyperbaric chambers or improperly calibrated scuba equipment can create conditions that may lead to convulsions if pure oxygen is breathed for greater than three hours. Breathing air at a depth of around 300 feet can be toxic and is equivalent to breathing pure oxygen at a depth of 66 feet.

At high pressures of oxygen, around 2,000 to 5,000 mm Hg, dangerous cerebral problems such as dizziness, twitching, vision deterioration, and nausea may occur. Continued exposure to these high pressures will result in confusion, convulsion, and eventual death.

Changes in total pressure can induce trapped gas effects. With a decrease in pressure, trapped gases will increase in volume (according to Boyle's law). Trapped gases in the body include air pockets in the ears, sinuses, and chest. Divers refer to the trapped gas phenomenon as the *squeeze*. Jet travel causes the most commonly occurring instance of trapped gas effects. Takeoff and landing may cause relatively sudden shifts in pressure, which may lead to discomfort and pain. With very rapid ascent or descent, injury can develop.

Lung rupture can be caused by a swift return to the surface from diving or decompression during high-altitude flight. This event is rare and happens only if the person is holding his or her breath during the decompression.

Evolved gas effects are associated with the absorption of nitrogen into body tissues. When breathed, nitrogen can be absorbed into all body tissues in concentrations proportional to the partial pressure of nitrogen in air. When a person is ascending in altitude, on the ground, in flight, or under water, nitrogen must be exhaled at a rate equal to or exceeding the absorption rate to avoid evolved gas effects.

If the nitrogen in body tissues such as blood is being absorbed faster than it is being exhaled, bubbles of gas may form in the blood and other tissues. Gas bubbles in the tissues may cause decompression sickness, which can be painful and occasionally fatal. Early symptoms of this disorder occur in body bends or joints such as elbows, knees, and shoulders. The common name for decompression sickness is the bends.

When the formation of gas bubbles is due to rapid ambient pressure reduction, it is called dysbarism.[23] The major causes of dysbarism are (1) the release of gas from the blood, and (2) the attempted expansion of trapped gas in body tissues. The sickness may occur with the decompression associated with rapidly moving from sea level (considered zero) to approximately 20,000 feet above sea level. Dysbarism is most often associated with underwater diving or working in pressurized containers (such as airplanes). Obese and older people seem to be more susceptible to dysbarism and decompression sickness.

Dysbarism manifests itself in a variety of symptoms. The creeps are caused by bubble formation in the skin, which causes an itchy, crawling, rashy feeling in the skin. Coughing and choking, resulting from bubbles in the respiratory system, is called the chokes. Bubbles occurring in the brain, although rare, may cause tingling and numbing, severe headaches, spasticity of muscles, and in some cases, blindness and paralysis. Dysbarism of the brain is rare. Rapid pressure change may also cause pain in the teeth and sinuses.[24]

Aseptic necrosis of bone is a delayed effect of decompression sickness. Bubbles in the capillaries supplying the bone marrow may become blocked with gas bubbles, which can cause a collection of platelets and blood cells to build up in a bone cavity. The marrow generation of blood cells can be damaged as well as the maintenance of healthy bone cells. Some bone areas may become calcified with severe complications when the bone is involved in a joint.

MEASUREMENT OF PRESSURE HAZARDS

Confirming the point of pressurized gas leakage can be difficult. After a gas has leaked out to a level of equilibrium with its surrounding air, the symptoms of the leak may disappear. There are several methods of detecting leaks (see Figure 8–9).

There are many potential causes of gas leaks. The most common of these are as follows:

- *Contamination* by dirt can prevent the proper closing of gas valves, threads, gaskets, and other closures used to control gas flow.
- *Overpressurization* can overstress the gas vessel, permitting gas release. The container closure may distort and separate from gaskets, leading to cracking.
- *Excessive temperatures* applied to dissimilar metals that are joined may cause unequal thermal expansion, loosening the metal-to-metal joint and allowing gas to escape. Materials may crack because of excessive cold, which may also result in gas escape. Thermometers are often used to indicate the possibility of gas release.
- *Operator errors* may lead to hazardous gas release from improper closure of valves, inappropriate opening of valves, or overfilling of vessels. Proper training and supervision can reduce operator errors.

Checklist for Detecting Leaks

✓ *Sounds* can be used to signal a pressurized gas leak. Gas discharge may be indicated by a whistling noise, particularly with highly pressurized gases escaping through small openings. Workers should not use their fingers to probe for gas leaks as highly pressurized gases may cut through tissue, including bone.

✓ *Cloth streamers* may be tied to the gas vessel to help indicate leaks. Soap solutions may be smeared over the vessel surface so that bubbles are formed when gas escapes. A stream of bubbles indicates gas release.

✓ *Scents* may be added to gases that do not naturally have an odor. The odor sometimes smelled in homes that cook or heat with natural gas is not the gas but a scent added to it.

✓ *Leak detectors* that measure pressure, current flow, or radioactivity may be useful for some types of gases.

✓ *Corrosion* may be the long-term effect of escaping gases. Metal cracking, surface roughening, and general weakening of materials may result from corrosion.

Figure 8-9
Convenient tool for detecting pressure leaks.

REDUCING PRESSURE HAZARDS

The reduction of pressure hazards often requires better maintenance and inspection of equipment that measures or uses high-pressure gases. Proper storage of pressurized containers reduces many pressure hazards. Pressurized vessels should be stored in locations away from cold or heat sources, including the sun. Cryogenic compounds (those that have been cooled to unusually low temperatures) may boil and burst the container when not kept at the proper temperatures. The whipping action of pressurized flexible hoses can also be dangerous. Hoses should be firmly clamped at the ends when pressurized.

Gas compression can occur in sealed containers exposed to heat. For this reason, aerosol cans must never be thrown into or exposed to a fire. Aerosol cans may explode violently when exposed to heat, although most commercially available aerosols are contained in low-melting point metals that melt before pressure can build up.

Pressure should be released before working on equipment. Gauges can be checked before any work on the pressurized system is begun. When steam equipment is shut down, liquid may condense within the system. This liquid and/or dirt in the system may become a propellant, which may strike bends in the system, causing loud noises and possible damage.

Water hammer is a series of loud noises caused by liquid flow suddenly stopping.[25] The momentum of the liquid is conducted back upstream in a shock wave. Pipe fittings and valves may be damaged by the shock wave. Reduction of this hazard involves using air chambers in the system and avoiding the use of quick-closing valves.

Negative pressures or vacuums are caused by pressures below atmospheric level. Negative pressures may result from hurricanes and tornadoes. Vacuums may cause

---

Checklist of Strategies for Reducing Pressure Hazards

✓ Install valves so that failure of the valve does not result in a hazard.

✓ Do not store pressurized containers near heat or sources of ignition.

✓ Train and test personnel dealing with pressurized vessels. Only tested personnel should be permitted to install, operate, maintain, calibrate, or repair pressurized systems. Personnel working on pressure systems should wear safety face shields or goggles.

✓ Examine valves periodically to ensure that they are capable of withstanding working pressures.

✓ Operate pressure systems only under the conditions for which they were designed.

✓ Relieve all pressure from the system before performing any work.

✓ Label pressure system components to indicate inspection status as well as acceptable pressures and flow direction.

✓ Connect pressure relief devices to pressure lines.

✓ Do not use pressure systems and hoses at pressure exceeding the manufacturer's recommendations.

✓ Keep pressure systems clean.

✓ Keep pressurized hoses as short as possible.

✓ Avoid banging, dropping, or striking pressurized containers.

✓ Secure pressurized cylinders by a chain to prevent toppling.

✓ Store acetylene containers upright.

✓ Examine labels before using pressurized systems to ensure correct matching of gases and uses.

Figure 8–10
Reduction of pressure hazards.

collapse of closed containers. Building code specifications usually allow for a pressure differential. Vessel wall thickness must be designed to sustain the load imposed by the differential in pressure caused by negative pressure. Figure 8–10 describes several methods to reduce the hazards associated with pressurized containers.

ANALYZING THE WORKPLACE

Regardless of whether the concern is temperature or pressure hazards, it is important to be able to conduct a thorough analysis of the workplace to identify hazards. The checklist in Figure 8–11 is a helpful tool for doing this. The first set of questions in the checklist pertains to temperature hazards. The second set pertains to pressure hazards.

Checklist for Analyzing the Workplace for Temperature and Pressure Hazards

Temperature Hazards

✓ Are workers in hot environments gradually acclimatized?

✓ Are workers in hot environments rotated into cooler environments at specified intervals?

✓ Do workers in hot environments wear personal protective clothing?

✓ Are first aid stations and supplies readily available for the treatment of burn victims?

✓ Are eye wash and emergency shower stations readily available for chemical burn victims?

✓ Are central or spot heating furnaces, warm air jets, contact warm plates, or radiant heaters used where employees work in cold environments?

✓ Are work areas in cold environments covered with insulating material?

✓ Are the handles of tools used in cold environments covered with insulating material?

✓ Are metal chairs in cold environments covered with insulating material?

✓ Are heated tents or shelters provided so that workers in cold environments can take periodic warming breaks?

✓ Are appropriate types of warm drinks made available to workers in cold environments?

✓ Are jobs performed in cold environments designed to require movement and minimize sitting or standing still?

✓ Are jobs performed in cold environments designed so that as many tasks as possible are performed in a warm environment?

Pressure Hazards

✓ Are boilers properly installed (level, sufficient room around them for inspecting all sides, and so on?)

✓ Are control/safety devices on boilers present and in proper working condition?

✓ Is a schedule of regular inspections for all boilers posted and clearly visible?

✓ Is a schedule of preventative maintenance for all boilers posted and adhered to?

✓ Are pulse-dampening devices/strategies being employed on high-pressure systems?

✓ Are high-pressure systems installed with a minimum of joints?

✓ Are appropriate pressure gauges in place and working properly?

✓ Are shields placed around high-pressure systems?

✓ Are remote and monitoring devices used with high-pressure systems?

✓ Is access restricted in areas where high-pressure systems are present?

✓ Are leak-detection methods being employed with pressured gas systems?

Figure 8-11
Convenient tool for preventing temperature and pressure hazards.

This checklist should be used periodically because conditions can change if new equipment is installed, personnel are replaced, or facility adaptations are made. Workplace analysis is a critical, ongoing responsibility.

APPLICATION SCENARIOS

1. An employee has passed out while working in conditions of high heat and humidity. His skin is hot, dry, red, and mottled. A call has been made for medical help. In the meantime, what should be done to help this employee?
2. Your company has experienced an unusual number of heat exhaustion cases. What can you recommend to reverse this dangerous trend?
3. An employee has been severely burned on the job. In completing the accident report, you are asked to estimate the percent of BSA burned. The burns covered the employee's back and neck. What percent of BSA should you enter in the report?
4. You have been asked to develop a program for preventing cold stress. Write out all aspects of the program.
5. Your company has added a new process that uses pressurized gas. You have been asked to suggest several methods for detecting gas leaks. What will you suggest?
6. Select a workplace to which you have access and apply the checklist in Figure 8-11. What potential problems can you identify?

ENDNOTES

1. Alpaugh, E. L. (Revised by T. J. Hogan) *Fundamentals of Industrial Hygiene,* 3rd ed. (Chicago: National Safety Council, 1988) pp. 259–260.
2. North, C. "Heat Stress," *Safety & Health,* April 1991, Vol. 141, No. 4, p. 55.
3. Hildebrand, J. E. "Radiation: Nuclear Power Industry Poses Unique Problems," *Occupational Health & Safety,* May 1987, Vol. 56, No. 5, p. 28.
4. Blocker, T. G., Jr. *Studies on Burns and Wound Healing* (Austin: University of Texas, 1965), p. 63.
5. National Safety Council, "Chemical Burns," Data Sheet 1–523 Rev. 87, p. 1.
6. Ibid., pp. 3–4.
7. Ibid., p. 2.
8. Ibid.
9. Alpaugh, E. L. *Fundamentals of Industrial Hygiene,* p. 261.
10. Ibid., p. 262.
11. Ibid.
12. Ibid., p. 261.
13. Ibid.
14. Ibid., p. 264.
15. Fraser, T. M. *The Worker at Work* (New York: Taylor & Francis, 1989), p. 300.
16. National Safety Council. *Accident Prevention Manual for Industrial Operations: Engineering and Technology,* 9th ed. (Chicago: National Safety Council, 1988), p. 485.
17. Ibid., p. 484.
18. Ibid., p. 488.
19. Ibid., pp. 487–89.
20. Ibid., p. 489.
21. Ibid.
22. Fraser, T. M. *The Worker at Work,* p. 315.
23. Hammer, W. *Occupational Safety Management and Engineering* (Upper Saddle River, NJ: Prentice Hall, 1989), p. 336.
24. Ibid., p. 339.
25. Ibid., p. 329.

Electrical and Fire Hazards

ELECTRICAL HAZARDS DEFINED

Electricity is the flow of negatively charged particles called electrons through an electrically conductive material. Electrons orbit the nucleus of an atom, which is located approximately in the atom's center. The negative charge of the electrons is neutralized by particles called neutrons, which act as temporary energy repositories for the interactions between positively charged particles called protons and electrons.

Electrons that are freed from an atom and are directed by external forces to travel in a specific direction produce electrical current, also called *electricity*. Conductors are substances that have many free electrons at room temperature and can pass electricity. Insulators do not have a large number of free electrons at room temperature and do not conduct electricity. Substances that are neither conductors nor insulators can be called semiconductors.

Electrical current passing through the human body causes a shock. The quantity and path of this current determines the level of damage to the body. The path of this flow of electrons is from a negative source to a positive point, since opposite charges attract one another.

When a surplus or deficiency of electrons on the surface of a material exists, static electricity is produced. This type of electricity is named "static" because there is no positive material nearby to attract the electrons and cause them to move. Friction is not required to produce static electricity, although it can increase the charge of existing static electricity. When two surfaces of opposite static electricity charges are brought into close range, a discharge, or spark, will occur. The spark from static electricity is often the first clue that such static exists. A common example is the sparks that come from rustling woolen blankets in dry heated indoor air.

The potential difference between two points in a circuit is measured by voltage. The higher the voltage, the more likely it is that electricity will flow between the negative and positive points.

Pure conductors offer little resistance to the flow of electrons. Insulators, on the other hand, have very high resistance to electricity. Semiconductors have a medium-range resistance to electricity. The higher the resistance, the lower the flow of electrons. Resistance is measured in ohms.

Electrical current is produced by the flow of electrons. The unit of measurement for current is amperes or amps. One amp is a current flow of 6.28×10^{18} electrons per second. Current is usually designated by I. Ohm's law describes the relationship among volts, ohms, and amps. One ohm is the resistance of a conductor that has a current of one amp under the potential of one volt. Ohm's law is stated as

$$V = IR$$

where

V = potential difference in volts
I = current flow in amps
R = resistance to current flow in ohms

Power is measured in wattage, or watts, and can be determined from Ohm's law:

$$W = VI \text{ or } W = I^2R$$

where

W = power in watts

Most industrial and domestic use of electricity is supplied by alternating current, or AC current. In the United States, standard AC circuits cycle 60 times per second. The number of cycles per second is known as frequency and is measured in hertz.

Electrical hazards occur when a person makes contact with a conductor carrying a current and simultaneously contacts the ground or another object that includes a conductive path to the ground. This person completes the circuit loop by providing a load for the circuit and thereby enables the current to pass through his or her body. People can be protected from this danger by insulating the conductors, insulating the people, or isolating the danger from the people.

SOURCES OF ELECTRICAL HAZARDS

Short circuits are one of many potential electrical hazards that can cause electrical shock. Another hazard is water, which considerably decreases the resistivity of materials, including humans. The resistance of wet skin can be around 450 ohms, whereas dry skin may have an average resistance of 100,000 ohms. According to Ohm's law, the higher the resistance, the lower the current flow. When the current flow is reduced, the probability of electrical shock is also reduced.

The major causes of electrical shock are shown in Figure 9–1.

Electrostatic Hazards

Electrostatic hazards may cause minor shocks. Shocks from static electricity may result from a single discharge or multiple discharges of static. Sources of electrostatic discharge are shown in Figure 9–2.

The rate of discharge of electrical charges increases with lower humidity. Electrostatic sparks are often greater during cold, dry winter days. Adding humidity to the air is

Figure 9–1
Major causes of electrical shock.

> **Checklist of Major Causes of Electrical Shock**
>
> ✓ Contact with a bare wire carrying current. The bare wire may have deteriorated insulation or be normally bare.
>
> ✓ Working with electrical equipment that lacks the UL label for safety inspection.
>
> ✓ Electrical equipment that has not been properly grounded. Failure of the equipment can lead to short circuits.
>
> ✓ Working with electrical equipment on damp floors or other sources of wetness.
>
> ✓ Static electricity discharge.
>
> ✓ Using metal ladders to work on electrical equipment. These ladders can provide a direct line from the power source to the ground, again causing a shock.
>
> ✓ Working on electrical equipment without ensuring that the power has been shut off.
>
> ✓ Lightning strikes.

not commonly used to combat static discharge, however, because higher humidity may result in an uncomfortable working environment and adversely affect equipment.[1]

Arcs and Sparks Hazards

With close proximity of conductors or contact of conductors to complete a circuit, an electric arc can jump the air gap between the conductors and ignite combustible gases or dusts. When the electric arc is a discharge of static electricity, it may be called a spark. A spark or arc may involve relatively little or a great deal of power and is usually discharged into a small space.

Combustible and Explosive Materials

High currents through contaminated liquids may cause the contaminants to expand rapidly and explode. This situation is particularly dangerous with contaminated oil-filled

Figure 9–2
Sources of electrostatic discharge.
Source: Hammer, W. *Occupational Safety Management and Engineering* (Upper Saddle River, NJ: Prentice Hall, 1989), p. 367.

> **Checklist of Sources of Electrostatic Discharge**
>
> ✓ Briskly rubbing a nonconductive material over a stationary surface. One common example of this is scuffing shoes across a wool or nylon carpet. Multilayered clothing may also cause static sparks.
>
> ✓ Moving large sheets of plastic, which may discharge sparks.
>
> ✓ The explosion of organic and metallic dusts, which have occurred from static buildup in farm grain silos and mine shafts.
>
> ✓ Conveyor belts. Depending on their constituent material, they can rub the materials being transported and cause static sparks.
>
> ✓ Vehicle tires rolling across a road surface.
>
> ✓ Friction between a flowing liquid and a solid surface.

circuit breakers or transformers. A poor match between current or polarity and capacitors can cause an explosion. In each of these cases, the conductor is not capable of carrying a current of such high magnitude. Overheating from high currents can also lead to short circuits, which in turn may generate fires and/or explosions.

Lightning Hazards

Lightning is static charges from clouds following the path of least resistance to the earth, involving very high voltage and current. If this path to the earth involves humans, serious disability may result, including electrocution. Lightning may also damage airplanes from intracloud and cloud-to-cloud flashes. Electrical equipment and building structures are commonly subject to lightning hazards. Lightning tends to strike the tallest object on the earth below the clouds. A tree is a common natural path for lightning.

Improper Wiring

Improper wiring permits equipment to operate normally but can result in hazardous conditions. One common unsafe wire practice is to jump the ground wire to the neutral wire. In this case, the ground wire is actually connected to the neutral wire. Equipment usually operates in a customary way, but the hazard occurs when low voltages are generated on exposed parts of the equipment, such as the housing. If the neutral circuit becomes corroded or loose, the voltage on the ground wire increases to a dangerous level.

Improper wiring is another common wiring error. When the ground is connected improperly, the situation is referred to as open ground. Usually the equipment with this miswiring will operate normally. If a short occurs in the equipment circuitry without proper grounding, anyone touching that equipment may be severely shocked.

Checklist of Environments Toxic to Insulation

✓ Direct sunlight or other sources of ultraviolet light, which can induce gradual breakdown of plastic insulation material.

✓ Sparks or arcs from discharging static electricity, which can result in burned-through holes in insulation.

✓ Repeated exposure to elevated temperatures, which can produce slow but progressive degradation of insulation material.

✓ Abrasive surfaces, which can result in erosion of the material strength of the insulation.

✓ Substance incompatibility with the atmosphere around the insulation and the insulation material, which can induce chemical reactions. Such reactions may include oxidation or dehydration of the insulation and eventual breakdown.

✓ Animals such as rodents or insects chewing or eating the insulation material, leading to exposure of the circuit. Insects can also pack an enclosed area with their bodies so tightly that a short circuit occurs. This is a common occurrence with electrical systems near water, such as pump housings and television satellite dishes.

✓ Moisture and humidity being absorbed by the insulation material, which may result in the moisture on the insulation carrying a current.

Figure 9–3
Environments that are toxic to insulation.

Figure 9–4
Common types of
equipment failure.

> **Checklist of Common
> Types of Equipment Failure**
>
> ✓ Wet insulation can become a conductor
>   and cause an electrical shock.
>
> ✓ Portable tool defects can result in the
>   device's housing carrying an electric cur-
>   rent. Workers do not expect tool housings
>   to be charged and may be shocked when
>   they touch a charged tool housing.
>
> ✓ Broken power lines carry great amperage
>   and voltage and can cause severe disability.
>
> ✓ When equipment is not properly grounded
>   or insulated, an unshielded worker may
>   receive a substantial electrical shock.

With reversed polarity, the hot and neutral wires have been reversed. A worker who is not aware that the black lead (hot) and white lead (neutral) have been reversed could be injured or cause further confusion by connecting the circuit to another apparatus. If a short between the on/off switch and the load occurred, the equipment may run indefinitely, regardless of the switch position. In a reversed polarity light bulb socket, the screw threads become conductors.[2]

Insulation Failure

The degradation of insulation can cause a bare wire and resulting shock to anyone coming in contact with that wire. Most insulation failure is caused by environments toxic to insulation. These environments are shown in Figure 9–3.

Equipment Failure

There are several ways in which equipment failure can cause electrical shocks. Electrical equipment designers attempt to create devices that are explosion-proof, dust-ignition-proof, and spark-proof. Figure 9–4 shows some of the more common types of equipment failure.

ELECTRICAL HAZARDS TO HUMANS

The greatest danger to humans suffering electrical shock results from current flow. The voltage determines whether a particular person's natural resistance to current flow will be overcome. Skin resistance can vary between 100,000 ohms and 600,000 ohms, depending on skin moisture.[3] Some levels of current "freeze" a person to the conductor; the person cannot voluntarily release his or her grasp. Let-go current is the highest current level at which a person in contact with the conductor can release the grasp of the conductor. Figure 9–5 shows the relationship between amperage dosage and danger with a typical domestic 60-cycle AC current.

The severity of injury with electrical shock depends on the dosage of current, as shown in Figure 9–5, but also on the path taken through the body by the current. The path is influenced by the resistance of various parts of the body at the time of contact with the conductor. The skin is the major form of resistance to current flow. Current paths through the heart, brain, or trunk are generally much more injurious than paths through extremities.

Checklist of Effects of Amperage Dosage	
Dose in Current in Milliamps	**Effect on Human Body**
✓ Less than 1	No sensation, no perceptible effect.
✓ 1	Shock perceptible, reflex action to jump away. No direct danger from shock but sudden motion may cause accident.
✓ More than 3	Painful shock.
✓ 6	Let-go current for women.
✓ 9	Let-go current for men.
✓ 10–15	Local muscle contractions. Freezing to the conductor for 2.5% of the population.
✓ 30–50	Local muscle contractions. Freezing to the conductor for 50% of the population.
✓ 50–100	Prolonged contact may cause collapse and unconsciousness. Death may occur after 3 minutes of contact due to paralysis of the respiratory muscles.
✓ 100–200	Contact of more than a quarter of a second may cause ventricular fibrillation of the heart and death. AC currents continuing for more than one heart cycle may cause fibrillation.
✓ More than 200	Clamps and stops the heart as long as the current flows. Heart beating and circulation may resume when current ceases. High current can produce respiratory paralysis, which can be reversed with immediate resuscitation. Severe burns to the skin and internal organs. May result in irreparable body damage.

Figure 9–5
Current effects on the human body.

REDUCING ELECTRICAL HAZARDS

Grounding of electrical equipment is the primary method of reducing electrical hazards. The purpose of grounding is to safeguard people from electrical shocks, reduce the probability of a fire, and protect equipment from damage. Grounding ensures a path to the earth for the flow of excess current. Grounding also eliminates the possibility of a person being shocked by contact with a charged capacitor. The actual mechanism of grounding was discussed at the beginning of this chapter.

Electrical system grounding is achieved when one conductor of the circuit is connected to the earth. Power surges and voltage changes are attenuated and usually eliminated with proper system grounding. Bonding is used to connect two pieces of equipment by a conductor. Bonding can reduce potential differences between the equipment and thus reduce the possibility of sparking. Grounding, in contrast, provides a conducting path between the equipment and the earth. Bonding and grounding together are used for entire electrical systems.

Separate equipment grounding involves connecting all metal frames of the equipment in a permanent and continuous manner. If an insulation failure occurs, the current should return to the system ground at the power supply for the circuit. The equipment ground wiring will be the path for the circuit current, enabling circuit breakers and fuses to operate properly. The exposed metal parts of the equipment shown in Figure 9–6 must be grounded or provided with double insulations.[4]

**Checklist of Equipment Requiring
Grounding or Double Insulation**

✓ Portable electric tools such as drills and saws.

✓ Communication receivers and transmitters.

✓ Electrical equipment in damp locations.

✓ Television antenna towers.

✓ Electrical equipment in flammable liquid storage areas.

✓ Electrical equipment operated with over 150 volts.

A ground fault interrupter (GFI), also called a ground fault circuit interrupter (GFCI), can detect the flow of current to the ground and open the circuit, thereby interrupting the flow of current. When the current flow in the hot wire is greater than the current in the neutral wire, a ground fault has occurred. The GFI provides a safety measure for a person who becomes part of the ground fault circuit. The GFI cannot interrupt current passing between two circuits or between the hot and neutral wires of a three-wire circuit. To ensure safety, equipment must be grounded as well as protected by a GFI.

There are several options for reducing the hazards associated with static electricity. The primary hazard of static electricity is the transfer of charges to surfaces with lower potential. Bonding and grounding are two means of controlling static discharge. Humidification is another mechanism for reducing electrical static; it was discussed in the section on sources of electrical hazards. Raising the humidity above 65 percent reduces charge accumulation.[5]

Anti-static materials have also been used effectively to reduce electrical static hazards. Such materials either increase the surface conductivity of the charged material or absorb moisture, which reduces resistance and the tendency to accumulate charges.

Ionizers and electrostatic neutralizers ionize the air surrounding a charged surface to provide a conductive path for the flow of charges. Radioactive neutralizers include a radioactive element that emits positive particles to neutralize collected negative electrical charges. Workers need to be safely isolated from the radioactive particle emitter.

Fuses consist of a metal strip or wire that will melt if a current above a specific value is conducted through the metal. Melting the metal causes the circuit to open at the fuse, thereby stopping the flow of current. Some fuses are designed to include a time lag before melting to allow higher currents during startup of the system or as an occasional event.

Magnetic circuit breakers use a solenoid, a type of coil, to surround a metal strip that connects to a tripping device. When the allowable current is exceeded, the magnetic force of the solenoid retracts the metal strip, opening the circuit. Thermal circuit breakers rely on excess current to produce heat and bending in a sensitive metal strip. Once bent, the metal strip opens the circuit. Circuit breakers differ from fuses in that they are usually easier to reset after tripping and often provide a lower time lag or none at all before being activated.

Double insulation is another means of increasing electrical equipment safety. Most double insulated tools have plastic nonconductive housings in addition to standard insulation around conductive materials.

There are numerous methods of reducing the risk of electrocution by lightning. Figure 9–7 lists the major precautions to take.[6]

Another means of protecting workers is isolating the hazard from the workers or vice versa. Interlocks automatically break the circuit when an unsafe situation is detected. Interlocks may be used around high-voltage areas to keep personnel from entering the area. Elevator doors typically have interlocks to ensure that the elevator does not move when the doors are open. Warning devices to alert personnel

---

Checklist of Strategies for Reducing Lightning Hazards

✓ Place lightning rods so that the upper end is higher than nearby structures.

✓ Avoid standing in high places or near tall objects. Be aware that trees in an open field may be the tallest object nearby.

✓ Do not work with flammable liquids or gases during electrical storms.

✓ Ensure proper grounding of all electrical equipment.

✓ If inside an automobile, remain inside the automobile.

✓ If in a small boat, lie down in the bottom of the boat.

✓ If in a metal building, stay in the building and do not touch the walls of the building.

✓ Wear rubber clothing if outdoors.

✓ Do not work touching or near conducting materials, especially those in contact with the earth such as fences.

✓ Avoid using the telephone during an electrical storm.

✓ Do not use electrical equipment during the storm.

✓ Avoid standing near open doors or windows where lightning may enter the building directly.

Figure 9–7
Lightning hazard control.

about detected hazards may include lights, colored indicators, on/off blinkers, audible signals, or labels.

It is better to design safety into the equipment and system than to rely on human behavior such as reading and following labels. Figure 9–8 summarizes the many methods of reducing electrical hazards.

OSHA'S ELECTRICAL STANDARDS

OSHA's standards relating to electricity are found in 29 C.F.R. 1910 (Subpart S). They are extracted from the National Electrical Code. This code should be referred to when more detail is needed than appears in OSHA's excerpts. Subpart S is divided into the following two categories of standards: (1) Design of Electrical Systems, and (2) Safety-Related Work Practices. The standards in each of these categories are as follows:

Design of Electrical Systems

1910.302	Electric utilization systems
1910.303	General requirements
1910.304	Wiring design and protection
1910.305	Wiring methods, components, and equipment for general use
1910.306	Specific-purpose equipment and installations
1910.307	Hazardous (classified) locations
1910.308	Special systems

Safety-Related Work Practices

1910.331	Scope
1910.332	Training
1910.333	Selection and use of work practices

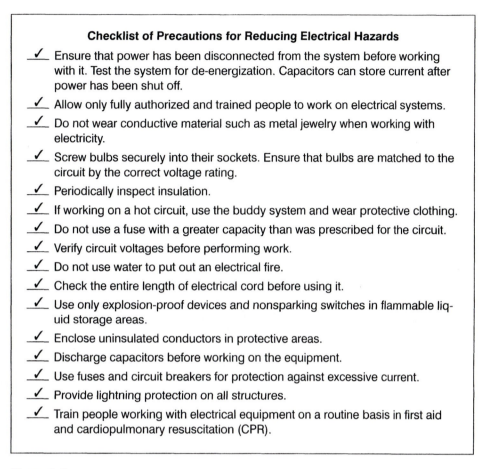

Checklist of Precautions for Reducing Electrical Hazards

✓ Ensure that power has been disconnected from the system before working with it. Test the system for de-energization. Capacitors can store current after power has been shut off.

✓ Allow only fully authorized and trained people to work on electrical systems.

✓ Do not wear conductive material such as metal jewelry when working with electricity.

✓ Screw bulbs securely into their sockets. Ensure that bulbs are matched to the circuit by the correct voltage rating.

✓ Periodically inspect insulation.

✓ If working on a hot circuit, use the buddy system and wear protective clothing.

✓ Do not use a fuse with a greater capacity than was prescribed for the circuit.

✓ Verify circuit voltages before performing work.

✓ Do not use water to put out an electrical fire.

✓ Check the entire length of electrical cord before using it.

✓ Use only explosion-proof devices and nonsparking switches in flammable liquid storage areas.

✓ Enclose uninsulated conductors in protective areas.

✓ Discharge capacitors before working on the equipment.

✓ Use fuses and circuit breakers for protection against excessive current.

✓ Provide lightning protection on all structures.

✓ Train people working with electrical equipment on a routine basis in first aid and cardiopulmonary resuscitation (CPR).

Figure 9–8
Summary of safety precautions for electrical hazards.

| 1910.334 | Use of equipment |
| 1910.335 | Safeguards for personal protection |

FIRE HAZARDS DEFINED

Fire hazards are conditions that favor fire development or growth. Three elements are required to start and sustain fire: (1) oxygen, (2) fuel, and (3) heat. Since oxygen is naturally present in most earth environments, fire hazards usually involve the mishandling of fuel or heat.

Fire, or combustion, is a chemical reaction between oxygen and a combustible fuel. Combustion is the process by which fire converts fuel and oxygen into energy, usually in the form of heat. By-products of combustion include light and smoke. For the reaction to start, a source of ignition, such as a spark or open flame, or a sufficiently high temperature is needed. Given a sufficiently high temperature, almost every substance will burn. The ignition temperature or combustion point is the temperature at which a given fuel can burst into flame.

Fire is a chain reaction. For combustion to continue, there must be a constant source of fuel, oxygen, and heat. Exothermic chemical reactions create heat. Combustion and fire are exothermic reactions and can often generate large quantities of heat. Endothermic reactions consume more heat than they generate. An ongoing fire usually provides its own sources of heat. It is important to remember that cooling is one of the principal ways to control a fire or put it out.

SOURCES OF FIRE HAZARDS

Almost everything in a workplace will burn. Metal furniture, machines, plaster, and concrete block walls are usually painted. Most paints and lacquers will easily catch fire. Oxygen is almost always present. Therefore, the principal method of fire suppression is passive—the absence of sufficient heat. Within our environment, various conditions elevate the risk of fire and so are termed *fire hazards*.

For identification, fires are classified according to their properties, which relate to the nature of the fuel. The properties of the fuel directly correspond to the best means of combating a fire (see Figure 9–9).

Without a source of fuel, there is no fire hazard. However, almost everything in our environment could be a fuel. Fuels occur as solids, liquids, vapors, and gases.

Solid fuels include wood, building decorations and furnishings such as fabric curtains and wall coverings, and synthetics used in furniture. What would an office be without paper? What would most factories be without cardboard and packing materials such as Styrofoam molds and panels, shredded or crumpled papers, bubble wrap, and shrink wrap? All of these materials easily burn.

Few solid fuels are or can be made fireproof. Even fire walls do not stop fires, although they are defined by their ability to slow the spread of fire. Wood and textiles can be treated with fire- or flame-retardant chemicals to reduce their flammability.

Solid fuels are involved in most industrial fires, but mishandling flammable liquids and flammable gases is a major cause of industrial fires. Two often-confused terms applied to flammable liquids are *flash point* and *fire point*. The flash point is the lowest temperature for a given fuel at which vapors are produced in sufficient concentrations to flash in the presence of a source of ignition. The fire point is the minimum temperature at which the vapors will continue to burn, given a source of ignition. The auto-ignition temperature is the lowest point at which the vapors of a liquid or solid will self-ignite *without* a source of ignition.

Flammable liquids have a flash point below 100°F. Combustible liquids have a flash point at or higher than 100°F. Both flammable and combustible liquids are further divided into the three classifications shown in Figure 9–10.

As the temperature of any flammable liquid increases, the amount of vapor generated on the surface also increases. Safe handling, therefore, requires both a knowledge of the properties of the liquid and an awareness of ambient temperatures in the work or storage place. The flammable range, or explosive range, defines the concentrations of a

Checklist of Classes of Fire	
✓ Class A fires	Solid materials such as wood, plastic, textiles, and their products: paper, housing, clothing.
✓ Class B fires	Flammable liquids and gases.
✓ Class C fires	Electrical (referring to live electricity situations, not including fires in other materials started by electricity).
✓ Class D fires	Combustible, easily oxidized metals such as aluminum, magnesium, titanium, and zirconium.
✓ Special categories	Extremely active oxidizers or mixtures, flammables containing oxygen, nitric acid, hydrogen peroxide, solid missile propellants.

Figure 9–9
Classes of fire.

Checklist of Classes of Flammable and Combustible Liquids

Flammable Liquids

✓ Class I–A   Flash point below 73°F, boiling point below 100°F.

✓ Class I–B   Flash point below 73°F, boiling point at or above 100°F.

✓ Class I–C   Flash point at or above 73°F, but below 100°F.

Combustible Liquids

✓ Class II   Flash point at or above 100°F, but below 140°F.

✓ Class III–A   Flash point at or above 140°F, but below 200°F.

✓ Class III–B   Flash point at or above 200°F.

Figure 9–10

Classes of flammable and combustible liquids.

Source: NFPA, *Fire Protection Handbook,* 15th ed. (Quincy, MA: National Fire Protection Association, 1981).

vapor or gas in air that can ignite from a source. The auto-ignition temperature is the lowest temperature at which liquids spontaneously ignite.

Most flammable liquids are lighter than water. If the flammable liquid is lighter than water, water cannot be used to put the fire out.[7] The application of water floats the fuel and spreads a gasoline fire. Crude oil fires will burn even while floating on fresh or sea water.

Unlike solids, which have a definite shape and location, and unlike liquids, which have a definite volume and are heavier than air, gases have no shape. Gases expand to fill the volume of the container in which they are enclosed, and they are frequently lighter than air. Released into air, gas concentrations are difficult to monitor due to the changing factors of air, current direction, and temperature. Gases may stratify in layers of differing concentrations but will often collect near the top of whatever container in which they are enclosed. Concentrations may have been sampled as being safe at workbench level but, at the same time, be close to or exceed flammability limits just above head height.

The products of combustion are gases, flame (light), heat, and smoke. Smoke is a combination of gases, air, and suspended particles, which are the products of incomplete combustion. Many of the gases present in smoke and at a fire site are toxic to humans. Other, usually nontoxic, gases may replace the oxygen normally present in air. Most fatalities associated with fire are from breathing toxic gases and smoke and from being suffocated because of oxygen deprivation. Gases that may be produced by a fire include acrolein, ammonia, carbon monoxide, carbon dioxide, hydrogen bromide, hydrogen cyanide, hydrogen chloride, hydrogen sulfide, sulfur dioxide, and nitrogen dioxide. Released gases are capable of traveling across a room and randomly finding a spark, flame, or adequate heat source, flashing back to the source of the gas.

The National Fire Protection Association (NFPA) has devised a system, NFPA 704, for the quick identification of hazards presented when substances burn. The NFPA's red, blue, yellow, and white diamond is used on product labels, shipping cartons, and buildings. Ratings within each category are 0 to 4, where zero represents no hazard; 4, the most severe hazard level. The colors refer to a specific category of hazard:

   Red = flammability
   Blue = health
   Yellow = reactivity
   White = special information

Figure 9–11 illustrates this hazard identification system.

Checklist of Fire Hazard Identification Color Code

Flammability has a red background and is the top quarter of the diamond.

0   No hazard. Materials are stable during a fire and do not react with water.

1   Slight hazard. Flash point well above normal ambient temperature.

2   Moderate hazard. Flash point is slightly above normal ambient temperature.

3   Extreme fire hazard. Gases or liquids that can ignite at normal temperature.

4   Extremely flammable gases or liquids with very low flash points.

Health has a blue background and is the left quarter of the diamond.

0   No threat to health.

1   Slight health hazards. Respirator is recommended.

2   Moderate health hazard. Respirator and eye protection required.

3   Extremely dangerous to health. Protective clothing and equipment is required.

4   Imminent danger to health. Breathing or skin absorption may cause death. A fully encapsulating suit is required.

Reactive has a yellow background and is the right quarter of the diamond.

0   No hazard. Material is stable in a fire and does not react with water.

1   Slight hazard. Materials can become unstable at higher temperatures, or react with water to produce a slight amount of heat.

2   Moderate or greater hazard. Materials may undergo violent chemical reaction, but will not explode; or materials that react violently with water directly or form explosive mixtures with water.

3   Extreme hazard. Materials may explode given an ignition source, or have violent reactions with water.

4   Constant extreme hazard. Materials may polymerize, decompose, explode, or undergo other hazardous reactions on their own. Area should be evacuated in event of a fire.

Special information has a white background and is the bottom quarter of the diamond.

This area is used to note any special hazards presented by the material.

Figure 9–11
Color code system for identification of fire hazards.

FIRE DANGERS TO HUMANS

Direct contact with flame is obviously dangerous to humans. Flesh burns, as do muscles and internal organs. The fact that we are 80 percent water, by some estimations, does not mitigate the fact that virtually all of the other 20 percent burns. Nevertheless, burns are not the major cause of death in a fire.

National Fire Protection Association statistics show that most people die in fires from suffocating or breathing smoke and toxic fumes. Carbon dioxide can lead to suffocation because it can be produced in large volumes, depleting oxygen from the air. Many fire extinguishers use carbon dioxide because of its ability to starve the fire of oxygen while simultaneously cooling the fire. The number one killer in fires is carbon monoxide, which is produced in virtually all fires involving organic compounds. Carbon monoxide is produced in large volumes and can quickly reach lethal dosage concentrations.

Figure 9–12 shows the major chemical products of combustion. Other gases may be produced under some conditions. Not all of these gases are present at any particular fire site. Many of these compounds will further react with other substances often present at a

Checklist of Chemical Products of Combustion		
Product	**Fuels**	**Pathology**
Acrolein	Cellulose, fatty substances, woods and paints	Highly toxic irritant to eyes and respiratory system.
Ammonia (NH_3)	Wool, silk, nylon, melamine, refrigerants, hydrogen-nitrogen compounds	Somewhat toxic irritant to eyes and respiratory system.
Carbon dioxide (CO_2)	All carbon and organic compounds	Not toxic, but depletes available oxygen.
Carbon monoxide (CO)	All carbon and organic compounds	Can be deadly.
Hydrogen chloride (HCN)	Wool, silk, nylon, paper, polyurethane, rubber, leather, plastic, wood	Quickly lethal asphyxiant.
Hydrogen sulfide (H_2S)	Sulfur-containing compounds, rubber, crude oil	Highly toxic gas. Strong odor of rotten eggs, but quickly destroys sense of smell.
Nitrogen dioxide (NO_2)	Cellulose nitrate, celluloid, textiles, other nitrogen oxides	Lung irritant, causing death or damage.
Sulfur dioxide (SO_2)	Sulfur and sulfur-containing compounds	Toxic irritant.

Figure 9–12
Major chemical products of combustion.

fire. For example, sulfur dioxide will combine with water to produce sulfuric acid. Oxides of nitrogen may combine with water to produce nitric acid. Sulfuric acid and nitric acid can cause serious acid burns.

REDUCING FIRE HAZARDS

The best way to reduce fires is to prevent their occurrence. A major cause of industrial fires is hot, poorly insulated machinery and processes. One means of reducing a fire hazard is the isolation of the three triangle elements: fuel, oxygen, and heat. In the case of fluids, closing a valve may stop the fuel element.

Fires may also be prevented by the proper storage of flammable liquids. Liquids should be stored as follows:

■ In flame-resistant buildings that are isolated from places where people work. Proper drainage and venting should be provided for such buildings.

■ In tanks below ground level.

■ On the first floor of multistory buildings.

Substituting less flammable materials is another effective technique for fire reduction. A catalyst or fire inhibitor can be employed to create an endothermic energy state that eventually will smother the fire. Several ignition sources can be eliminated or isolated from fuels:

■ Smoking should be prohibited near any possible fuels.

■ Electrical sparks from equipment, wiring, or lightning should not be close to fuels.

■ Open flames should be kept separate from fuels. These may include welding torches, heating elements, or furnaces.

- Tools or equipment that may produce mechanical or static sparks must also be isolated from fuels.

Other strategies for reducing the risk of fires are as follows:

- Clean up spills of flammable liquids as soon as they occur, and properly dispose of the materials used in the clean-up.
- Keep work areas free from extra supplies of flammable materials (e.g., paper, rags, boxes, and so on). Have only what is needed on hand with the remaining inventory properly stored.
- Run electrical cords along walls rather than across aisles or in other trafficked areas. Cords that are walked on can become frayed and dangerous.
- Turn off the power and completely de-energize equipment before conducting maintenance procedures.
- Don't use spark- or friction-prone tools near combustible materials.
- Routinely test fire extinguishers.

Fire Extinguishing Systems

Automatic sprinkler systems are an example of a fixed extinguishing system since the sprinklers are fixed in position. Water is the most common fluid released from the sprinklers. Sprinkler supply pipes may be kept filled with water in heated buildings; in warmer climates, valves are used to fill the pipes with water when the sprinklers are activated. When a predetermined heat threshold is breached, water flows to the heads and is released from the sprinklers.

Portable fire extinguishers are classified by the types of fire that they can most effectively reduce. Figure 9–13 describes the four major fire extinguisher classifications. Blocking or shielding the spread of fire can include covering the fire with an inert foam, inert powder, nonflammable gas, or water with a thickening agent added. The fire may

Checklist of Fire Extinguisher Characteristics			
Fire Class	Extinguisher Contents	Mechanism	Disadvantages
A	Foam, water, dry chemical	Cooling, smothering, dilution, breaks the fire, reaction chain	Freezing if not kept heated
B	Dry chemical, bromotrifluoromethane, and other halogenated compounds, foam, CO_2, dry chemical	Chain-breaking smothering, cooling, shielding	Halogenated compounds are toxic
C	Bromotrifluoromethane, CO_2, dry chemical	Chain-breaking smothering, cooling, shielding	Halogenated compounds are toxic; fires may ignite after CO_2 dissipates
D	Specialized powders such as graphite, sand	Cooling, smothering	Expensive cover of powder may be broken with resultant reignition

Figure 9–13
Fire extinguisher characteristics.

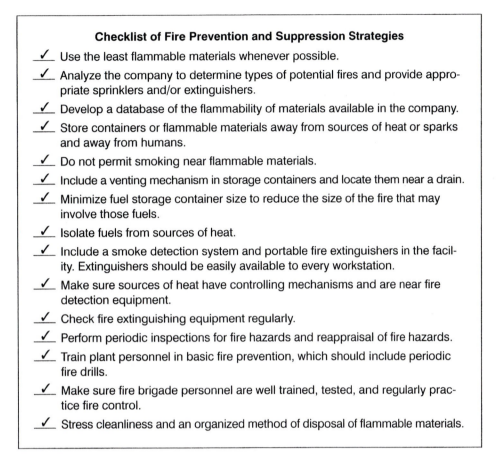

Checklist of Fire Prevention and Suppression Strategies

✓ Use the least flammable materials whenever possible.

✓ Analyze the company to determine types of potential fires and provide appropriate sprinklers and/or extinguishers.

✓ Develop a database of the flammability of materials available in the company.

✓ Store containers or flammable materials away from sources of heat or sparks and away from humans.

✓ Do not permit smoking near flammable materials.

✓ Include a venting mechanism in storage containers and locate them near a drain.

✓ Minimize fuel storage container size to reduce the size of the fire that may involve those fuels.

✓ Isolate fuels from sources of heat.

✓ Include a smoke detection system and portable fire extinguishers in the facility. Extinguishers should be easily available to every workstation.

✓ Make sure sources of heat have controlling mechanisms and are near fire detection equipment.

✓ Check fire extinguishing equipment regularly.

✓ Perform periodic inspections for fire hazards and reappraisal of fire hazards.

✓ Train plant personnel in basic fire prevention, which should include periodic fire drills.

✓ Make sure fire brigade personnel are well trained, tested, and regularly practice fire control.

✓ Stress cleanliness and an organized method of disposal of flammable materials.

Figure 9–14
Fire prevention and suppression summary.

suffocate under such a covering. Flooding a liquid fuel with nonflammable liquid can dilute this fire element.

Preventing Office Fires

The shop floor is not the only part of the plant where fire hazards exist. Offices are also susceptible to fires. According to Vogel, "Every year about 7,000 fires occur in office buildings, which cause injuries, deaths, and millions of dollars in fire damages."[8] The following strategies are helpful in preventing office fires:

■ Confine smoking to designated areas that are equipped with nontip ashtrays and fire-resistant furnishings.

■ Periodically check electrical circuits and connections. Replace frayed or worn cords immediately.

■ Make sure that extension cords and other accessories are UL approved and used only as recommended.

■ Make sure there is plenty of air space left around copying machines and other office machines that might overheat.

■ Locate heat-producing appliances away from the wall or anything else that could ignite.

■ Frequently inspect personal appliances such as hot plates, coffee pots, and cup warmers. Assign responsibility for turning off such appliances every day to a specific person.

■ Keep aisles, stairwells, and exits clear of paper, boxes, and other combustible materials.[9]

Figure 9–14 summarizes major fire prevention and suppression suggestions.

OSHA FIRE STANDARDS

OSHA standards for fire protection appear in 29 C.F.R. (Subpart L). This subpart contains the standards for fire brigades, fixed fire suppression equipment, and other fire protection systems. Employers are not required to form fire brigades, but those that choose to do so must meet a number of specific requirements. There are other fire-related requirements in other subparts. For example, fire exits, emergency action plans, and means of egress are covered in Subpart E. The standards in Subpart L are as follows:

Fire Protection

1910.155	Scope, application, and definitions
1910.156	Fire brigades

Portable Fire Suppression Equipment

1910.157	Portable fire extinguishers
1910.158	Standpipe and hose systems

Fixed Fire Suppression Equipment

1910.159	Automatic sprinkler systems
1910.160	Fixed extinguishing systems, general
1910.161	Fixed extinguishing systems, dry chemical
1910.162	Fixed extinguishing systems, gaseous agent
1910.163	Fixed extinguishing systems, water spray and foam

Other Fire Protection Systems

1910.164	Fire detection systems
1910.165	Employee alarm systems

LIFE SAFETY

Life safety involves protecting the vehicles, vessels, and lives of people in buildings and structures from fire. The primary reference source for life safety is the Life Safety Code, published by the National Fire Protection Association. The code applies to new and existing buildings. It addresses the construction, protection, and occupancy features necessary to minimize the hazards of fire, smoke fumes, and/or panic. A major part of the code is devoted to the minimum requirements for design of egress necessary to ensure that occupants can quickly evacuate a building or structure.

Basic Requirements[10]

In this section, the term *structure* refers to a structure or building.

- Every structure, new and existing, that is to be occupied by people must have a means of egress and other fire protection safeguards that, together, meet the following criteria: (1) ensure that occupants can promptly evacuate or be adequately protected without evacuating; and (2) provide sufficient backup safeguards to ensure that human life is not endangered if one system fails.

- Every structure must be constructed or renovated, maintained, and operated in such a way that occupants are: (1) protected from fire, smoke, or fumes; (2) protected from fire-related panic; (3) protected long enough to allow a reasonable amount of time for evacuation; and (4) protected long enough to defend themselves without evacuating.

■ In providing structures with means of egress and other fire protection safeguards, the following factors must be considered: (1) character of the occupancy; (2) capabilities of occupants; (3) number of occupants; (4) available fire protection; (5) height of the structure; (6) type of construction; and (7) any other applicable concerns.

■ No lock or other device may be allowed to obstruct egress in any part of a structure at any time that it is occupied. The only exceptions to this requirement are mental health detention and correctional facilities. In these, the following criteria are required: (1) responsible personnel must be available to act in the case of fire or a similar emergency; and (2) procedures must be in place to ensure that occupants are evacuated in the event of an emergency.

■ All exits in structures must satisfy the following criteria: (1) be clearly visible or marked in such a way that an unimpaired individual can readily discern the route of escape; (2) all routes to a place of safety must be arranged or clearly marked; (3) any doorway and passageway that might be mistaken as a route to safety must be arranged or clearly marked in such a way as to prevent confusion in an emergency; and (4) all appropriate steps must be taken to ensure that occupants do not mistakenly enter a dead-end passageway.

■ Egress routes and facilities must be included in the lighting design wherever artificial illumination is required in a structure.

Fire alarm systems must be provided in any facility that is large enough or so arranged that a fire itself might not adequately warn occupants of the danger. Fire alarms should alert occupants to initiate appropriate emergency procedures.

■ In any structure or portion of a structure in which a single means of egress might be blocked or overcrowded in an emergency situation, at least two means of egress must be provided. The two means of egress must be arranged in such a way as to minimize the possibility of both becoming impossible in the same emergency situation.

■ All stairs, ramps, and other means of moving from floor to floor must be enclosed (or otherwise protected) to afford occupants protection when used as a means of egress in an emergency situation. These means of vertical movement should also serve to inhibit the spread of fire, fumes, and smoke from floor to floor.

■ Compliance with the requirements summarized herein does not eliminate or reduce the need to take other precautions to protect occupants from fire hazards, nor does it permit the acceptance of any condition that could be hazardous under normal occupancy conditions.

The information in this section is a summary of the broad fundamental requirements of the *Life Safety Code* of the National Fire Protection Association. More specific requirements relating to means of egress and features of fire protection are explained in the sections that follow.

Means of Egress[11]

This section explains some of the more important issues in the *Life Safety Code* relating to means of egress. Students and practitioners who need more detailed information are encouraged to refer to the *Life Safety Code*. It is an invaluable resource for safety professionals.

1. *Doors.* Doors that serve as exits must be designed, constructed, and maintained in such a way that the means of egress is direct and obvious. Windows that could be mistaken for doors in an emergency situation must be made inaccessible to occupants.

2. *Capacity of means of egress.* The means of egress must have a capacity sufficient to accommodate the occupant load of the structure calculated in accordance with the requirements of the *Life Safety Code*.

3. *Number of means of egress.* Any component of a structure must have a minimum of two means of egress (with exceptions as set forth in the code). The minimum number of means of egress from any story or any part of a story is three for occupancy loads of 500 to 1,000 and four for occupancy loads of more than 1,000.

4. *Arrangement of means of egress.* All exits must be easily accessible at all times in terms of both location and arrangement.

5. *Measurement of travel distance to exits.* The travel distance to at least one exit must be measured on the walking surface along a natural path of travel beginning at the most remote occupied space and ending at the center of the exit. Distances must comply with the code.

6. *Discharge from exits.* All exits from a structure must terminate at a public way or at yards, courts, or open spaces that lead to the exterior of the structure.

7. *Illumination of means of egress.* All means of egress shall be illuminated continuously during times when the structure is occupied. Artificial lighting must be used as required to maintain the necessary level of illumination. Illumination must be arranged in such a way that no area is left in darkness by a single lighting failure.

8. *Emergency lighting.* Emergency lighting for all means of egress must be provided in accordance with the code. In cases where maintaining the required illumination depends on changing from one source of power to another, there shall be no appreciable interruption of lighting.

9. *Marking of means of egress.* Exits must be marked by readily visible, approved signs in all cases where the means of egress is not obviously apparent to occupants. No point in the exit access corridor shall be more than 100 feet from the nearest sign.

10. *Special provisions for high hazard areas.* If an area contains contents that are classified as highly hazardous, occupants must be able to exit by traveling no more than 75 feet. At least two means of egress must be provided, and there shall be no dead-end corridors.

The requirements summarized in this section relate to the fundamental specifications of the *Life Safety Code* relating to means of egress. For more detailed information concerning general requirements, means of egress, and other factors such as fire protection and fire protection equipment, refer to the actual code.

FLAME-RESISTANT CLOTHING[12]

For employees who work in jobs in which flames or electric arcs may occur, wearing flame-resistant clothing can be a life saver. Electric arcs are the result of electricity passing through ionized air. Although electric arcs last for only a few seconds, during that time they can produce extremely high levels of heat and flash flame.

OSHA's standards relating to flame-resistant clothing are found in C.F.R. 1910.269, paragraph 1. Key elements of paragraph 1 explain the employer's responsibilities regarding personal protective equipment and flame-resistant clothing.

Apparel[13]

C.F.R. 1910.269, paragraph 1(6) reads as follows:

(i)   When work is performed within reaching distance of exposed energized parts of equipment, the employer shall ensure that each employee removes or renders nonconductive all exposed conductive articles, such as key or watch chains, rings, or wrist watches or bands, unless such articles do not increase the hazards associated with contact with the energized parts.

(ii)   The employer shall train each employee who is exposed to the hazards or flames or electric arcs in the hazards involved.

(iii)   The employer shall ensure that each employee who is exposed to the hazards of flames or electric arcs does not wear clothing that, when exposed to flames or electric arcs, could increase the extent of injury that would be sustained by the employee. *Note:* Clothing made from the following types of fabrics, either alone or in blends, is prohibited by this paragraph, unless the employer can demonstrate that the fabric has been treated to withstand the conditions that may be encountered or that the clothing is worn in such a manner as to eliminate the hazard involved: acetate, nylon, polyester, rayon.

(iv) *Fuse Handling.* When fuses must be installed or removed with one or both terminals energized at more than 300 volts or with exposed parts energized at more than 50 volts, the employer shall ensure that tools or gloves rated for the voltage are used. When expulsion-type fuses are installed with one or both terminals energized at more than 300 volts, the employer shall ensure that each employee wears eye protection meeting the requirements of Subpart I of this Part, uses a tool rated for the voltage, and is clear of the exhaust path of the fuse barrel.

APPLICATION SCENARIOS

1. A fellow employee works with electricity all day and is careful to ensure that the insulation on wires is not broken or eroded before handling them. However, he never checks for moisture on the insulated wires. Is this a problem? Why or why not?

2. Locate a copy of the current OSHA standard for electricity. What are the requirements for safeguards for personal protection?

3. "It must be terrible to die in a fire," said George to his wife as they checked into the hotel. "That's why I always ask for a room near the elevator. I know if I'm near the elevator, I'll always have time to escape. In fact, as long as I don't see flames, I'll stay put and wait for rescue personnel. I'd rather breathe a little smoke than get burned by hot flames." Is George's assessment of his situation accurate? Why or why not?

4. A secretary in an office building is having trouble with her stapler, pens, pencils, and other supplies being taken. The copy room is separated from her office by a door that does not lock. As people make copies, they often "borrow" from her desk little things they need or have forgotten. Consequently, she has submitted a request to have maintenance personnel install a slide lock on her side of the door. This will allow her to keep people out of her office during the day but will still give her access to the copy machine. However, it will lock the second exit for people using the copy machine. Is this a problem? Why or why not?

ENDNOTES

1. Kavianian, H. R., and C. A. Wentz, Jr. *Industrial Safety* (New York: Van Nostrand Reinhold, 1990), p. 231.
2. Asfahl, C. Ray. *Industrial Safety and Health Management* (Upper Saddle River, NJ: Prentice Hall, 1990), p. 347.
3. Kavianian and Wentz, *Industrial Safety,* p. 214.
4. Ibid., p. 218.
5. Hammer, W. *Occupational Safety Management and Engineering* (Upper Saddle River, NJ: Prentice Hall), p. 372.
6. Kavianian and Wentz, *Industrial Safety,* p. 232.
7. Kavianian, H. R., and Wentz, C. A., Jr. *Fire Safety Manual* (New York: Van Nostrand Reinhold, 1990), p. 175.
8. Vogel, C. "Fires Can Raze Office Buildings," *Safety & Health,* September 1991, Vol. 144, No. 3, p. 27.
9. Ibid., pp. 26–27.
10. National Fire Protection Association. *Life Safety Code* (NFPA 101), 1997, pp. 101–19.
11. Ibid., pp. 101-26 through 101–50.
12. Neal, Thomas E. "The Need for Flame-Resistant Protective Clothing," *Occupational Safety,* May 1997, p. 96.
13. C.F.R. 1910.269, paragraph 1(6).

Toxic Substances and Confined Spaces

TOXIC SUBSTANCES DEFINED

A toxic substance is one that has a negative effect on the health of a person or animal. Toxic effects are a function of several factors, including the following: (1) properties of the substance, (2) amount of the dose, (3) level of exposure, (4) route of entry, and (5) resistance of the individual to the substance. Smith describes the issue of toxic substances as follows:

> When a toxic chemical acts on the human body, the nature and extent of the injurious response depends upon the dose received—that is, the amount of the chemical that actually enters the body or system and the time interval during which this dose was administered. Response can vary widely and might be as little as a cough or mild respiratory irritation or as serious as unconsciousness and death.[1]

ENTRY POINTS FOR TOXIC AGENTS

The development of preventive measures to protect against the hazards associated with industrial hygiene requires first knowing how toxic agents enter the body. A toxic substance must first enter the bloodstream to cause health problems. The most common routes of entry for toxic agents are inhalation, absorption, injection, and ingestion. These routes are explained in the following paragraphs.

Inhalation

The route of entry about which safety and health professionals should be most concerned is inhalation. Airborne toxic substances such as gases, vapors, dust, smoke, fumes, aerosols, and mists can be inhaled and pass through the nose, throat, bronchial tubes, and lungs to enter the bloodstream. The amount of a toxic substance that can be inhaled depends on the following factors: (1) concentration of the substance, (2) duration of exposure, and (3) breathing volume.

According to Olishifski,

> Inhalation, as a route of entry, is particularly important because of the rapidity with which a toxic material can be absorbed in the lungs, pass into the bloodstream, and reach the brain. Inhalation is the major route of entry for hazardous chemicals in the work environment.[2]

Absorption[3]

The second most common route of entry in an industrial setting is absorption, or passage through the skin and into the bloodstream. The human skin is a protective barrier against many hazards. However, certain toxic agents can penetrate the barrier through absorption. Of course, unprotected cuts, sores, and abrasions facilitate the process, but even healthy skin will absorb certain chemicals. Humans are especially susceptible to absorbing such chemicals as organic lead compounds, nitro compounds, organic phosphate pesticides, TNT, cyanides, aromatic amines, amides, and phenols.

With many substances, the rate of absorption and, in turn, the hazard levels increase in a warm environment. The extent to which a substance can be absorbed through the skin depends on the factors shown in Figure 10–1. Another factor is body site. Different parts of the body have different absorption capabilities. For example, the forearms have a lower absorption potential than do the scalp and forehead.

Ingestion[4]

Ingestion, not a major concern in an industrial setting, is entry through the mouth. An ingested substance is swallowed. It moves through the stomach into the intestines and from there into the bloodstream. Toxic agents sometimes enter the body by ingestion when workers eating lunch or a snack accidentally consume them. Airborne contaminants can also rest on food or the hands and, as a result, be ingested during a meal or snack. The possibility of ingesting toxic agents makes it critical to confine eating and drinking to sanitary areas away from the worksite and to make sure that workers practice good personal hygiene such as washing their hands thoroughly before eating or drinking.

As it moves through the gastrointestinal tract, the toxic substance's strength may be diluted. In addition, depending on the amount and toxicity of the substance, the liver may be able to convert it to a nontoxic substance. The liver can, at least, decrease the

Figure 10–1
Factors that affect absorption rates of toxic substances through the skin.

Checklist of Factors that Affect Absorption Rates

✓ Molecular size

✓ Degree of ionization

✓ Lipid solubility

✓ Aqueous solubility

level of toxicity and pass some of the substance along to the kidneys, where some of the substance is eliminated in the urine.

Injection

Injection involves introducing a substance into the body using a needle and syringe. Consequently, this is not often a route of entry for a toxic substance in the workplace. Injection is sometimes used for introducing toxic substances in experiments involving animals. However, this approach can produce misleading research results because the needle bypasses some of the body's natural protective mechanisms.

EFFECTS OF TOXIC SUBSTANCES

The effects of toxic substances vary widely, as do the substances themselves. However, all of the various effects and exposure times can be categorized as being either acute or chronic.

Acute effects/exposures involve a sudden dose of a highly concentrated substance. They are usually the result of an accident (a spill or damage to a pipe) that results in an immediate health problem ranging from irritation to death. Acute effects/exposures are (1) sudden, (2) severe, (3) typically involve just one incident, and (4) cause immediate health problems. Acute effects/exposures are not the result of an accumulation over time.

Chronic effects/exposures involve limited continual exposure over time. Consequently, the associated health problems develop slowly. The characteristics of chronic effects/exposures are (1) continual exposure over time, (2) limited concentrations of toxic substances, (3) progressive accumulation of toxic substances in the body and/or progressive worsening of associated health problems, and (4) little or no awareness of exposures on the part of affected workers.

When a toxic substance enters the body, it eventually affects one or more body organs. Part of the liver's function is to collect such substances, convert them to

Checklist of Toxic Substances and the Organs They Endanger			
Blood	**Kidneys**	**Heart**	**Brain**
✓ Benzene	✓ Mercury	✓ Aniline	✓ Lead
✓ Carbon monoxide	✓ Chloroform		✓ Mercury
✓ Arsenic			✓ Benzene
✓ Aniline			✓ Manganese
✓ Toluene			✓ Acetaldehyde
Eyes	**Skin**	**Lungs**	**Liver**
✓ Cresol	✓ Nickel	✓ Asbestos	✓ Chloroform
✓ Acrolein	✓ Phenol	✓ Chromium	✓ Carbon tetrachloride
✓ Benzyl chloride	✓ Trichloroethylene	✓ Hydrogen sulfide	✓ Toluene
✓ Butyl alcohol		✓ Mica	
		✓ Nitrogen dioxide	

Figure 10–2
Selected toxic substances and the organs that they endanger.

nontoxics, and send them to the kidneys for elimination in the urine. However, when the dose is more than the liver can handle, toxics move on to other organs, producing a variety of different effects. The organs that are affected by toxic substances are the blood, kidneys, heart, brain, central nervous system, skin, liver, lungs, and eyes. Figure 10–2 lists some of the more widely used toxic substances and the organs that they endanger most.

RELATIONSHIP OF DOSES AND RESPONSES

Safety and health professionals are interested in predictability when it comes to toxic substances. How much of a given substance is too much? What effect will a given dose of a given substance produce? These types of questions concern dose-response relationships. A dose of a toxic substance can be expressed in a number of different ways depending on the characteristics of the substance, for example, amount per unit of body weight, amount per body-surface area, or amount per unit of volume of air breathed. Smith expresses the dose-response relationship mathematically as follows:[5]

$$(C) \times (T) = K$$

where:

C = concentration
T = duration (time) of exposure
K = constant

Note that in this relationship, C times T is *approximately* equal to K. The relationship is not exact.

Three important concepts to understand relating to doses are dose threshold, lethal dose, and lethal concentration. These concepts are explained in the following paragraphs.

Dose Threshold

The dose threshold is the minimum dose required to produce a measurable effect. Of course, the threshold is different for different substances. In animal tests, thresholds are established using such methods as (1) observing pathological changes in body tissues, (2) observing growth rates (are they normal or retarded?), (3) measuring the level of food intake (has there been a loss of appetite?), and (4) weighing organs to establish body weight to organ weight ratios.

Lethal Dose

A lethal dose of a given substance is the dose that is highly likely to cause death. Such doses are established through experiments on animals. When lethal doses of a given substance are established, they are typically accompanied by information that is of value to medical professionals and industrial hygienists. Such information includes the type of animal used in establishing the lethal dose, how the dose was administered to the animal, and the duration of the administered dose. Lethal doses do not apply to inhaled substances. With these substances, the concept of lethal concentration is applied.

Lethal Concentration

A lethal concentration of an inhaled substance is the concentration that is highly likely to result in death. With inhaled substances, the duration of exposure is critical because the amount inhaled increases with every unprotected breath.

AIRBORNE CONTAMINANTS[6]

It is important to understand the different types of airborne contaminants that may be present in the workplace. Each type of contaminant has a specific definition that must be understood in order to develop effective safety and health measures to protect against it. The most common types of airborne contaminants are shown in Figure 10–3).

Dusts

Dusts are various types of solid particles that are produced when a given type of organic or inorganic material is scraped, sawed, ground, drilled, handled, heated, crushed, or otherwise deformed. The degree of hazard represented by dust depends on the toxicity of the parent material and the size and level of concentration of the particles.

Fumes

The most common causes of fumes in the workplace are such manufacturing processes as welding, heat treating, and metalizing, all of which involve the interaction of intense heat with a parent material. The heat volatilizes portions of the parent material, which then condenses as it comes in contact with cool air. The result of this reaction is the formation of tiny particles that can be inhaled.

Smoke

Smoke is the result of the incomplete combustion of carbonaceous materials. Because combustion is incomplete, tiny soot and/or carbon particles remain and can be inhaled.

Aerosols

Aerosols are liquid or solid particles that are so small they can remain suspended in air long enough to be transported over a distance. They can be inhaled.

Mists

Mists are tiny liquid droplets suspended in air. Mists are formed in two ways: when vapors return to a liquid state through condensation, and when the application of sudden force or pressure turns a liquid into particles.

Gases

Unlike other airborne contaminants that take the form of either tiny particles or droplets, gases are formless. Gases are actually formless fluids. Gases become particularly haz-

Figure 10–3
Common airborne
contaminants.

Checklist of Common Airborne Contaminants
✓ Dust
✓ Fumes
✓ Smoke
✓ Aerosols
✓ Mists
✓ Gases
✓ Vapors

ardous when they fill a confined, unventilated space. The most common sources of gases in an industrial setting are from welding and the exhaust from internal combustion engines.

Vapors

Certain materials that are solid or liquid at room temperature and at normal levels of pressure turn to vapors when heated or exposed to abnormal pressure. Evaporation is the most common process by which a liquid is transformed into a vapor.

In protecting workers from the hazards of airborne contaminants, it is important to know the permissible levels of exposure for a given contaminant and to monitor continually the level of contaminants using accepted measurement practices and technologies. The topic of exposure thresholds is covered later in this chapter.

EFFECTS OF AIRBORNE TOXICS

Airborne toxic substances are also classified according to the type of effect they have on the body. The primary classifications are irritants, asphyxiants, narcotics, and anesthetics. With all airborne contaminants, concentration and duration of exposure are critical concerns.

Irritants

Irritants are substances that cause irritation to the skin, eyes, and the inner lining of the nose, mouth, throat, and upper respiratory tract. However, they produce no irreversible damage. According to Smith,

> Irritants can be subdivided into primary and secondary irritants. A primary irritant is a material that exerts little systemic toxic action, either because the products formed on the tissues of the respiratory tract are nontoxic or because the irritant action is far in excess of any systemic toxic action. A secondary irritant produces irritant action on mucous membranes, but systemic effects resulting from absorption overshadow this effect. Normally, irritation is a completely reversible phenomenon.[7]

Asphyxiants

Asphyxiants are substances that can disrupt breathing so severely that suffocation results. Asphyxiants may be simple or chemical in nature. A simple asphyxiant is an inert gas that dilutes oxygen in the air to the point that the body cannot take in enough air to satisfy its needs for oxygen. Common simple asphyxiants include carbon dioxide, ethane, helium, hydrogen, methane, and nitrogen. Chemical asphyxiants, by chemical action, interfere with the passage of oxygen into the blood or the movement of oxygen from the lungs to body tissues. Either way, the end result is suffocation due to insufficient or no oxygenation. Common chemical asphyxiants include carbon monoxide, hydrogen cyanide, and hydrogen sulfide.

Narcotics/Anesthetics

Narcotics and anesthetics are similar in that carefully controlled dosages can inhibit the normal operation of the central nervous system without causing serious or irreversible effects. This makes them particularly valuable in a medical setting. Dentists and physicians use narcotics and anesthetics to control pain before and after surgery. However, if the concentration of the dose is too high, narcotics and anesthetics can cause unconsciousness and even death. When this happens, death is the result of asphyxiation. Widely used narcotics and anesthetics include acetone, methyl-ethyl-ketone, acetylene hydrocarbons, ether, and chloroform.

EFFECTS OF CARCINOGENS

A carcinogen is any substance that can cause a malignant tumor or a neoplastic growth. A *neoplasm* is cancerous tissue or tissue that might become cancerous. Other terms used synonymously for carcinogen are *tumorigen, oncogen,* and *blastomogen.*

According to Smith,

> It is well established that exposure to some chemicals can produce cancer in laboratory animals and man. There are a number of factors that have been related to the incidence of cancer—the genetic pattern of the host, viruses, radiation including sunshine, and hormone imbalance, along with exposure to certain chemicals. Other factors such as cocarcinogens and tumor accelerators are involved. It is also possible that some combination of factors must be present to induce cancers. There is pretty good clinical evidence that some cancers are virus-related. It may be that a given chemical in some way inactivates a virus, activates one, or acts as a cofactor.[8]

Medical researchers are not sure exactly how certain chemicals cause cancer. However, there are a number of toxic substances that are either known, or are strongly suspected, to be carcinogens. These include coal tar, pitch, creosote oil, anthracene oil, soot, lamp black, lignite, asphalt, bitumen waxes, paraffin oils, arsenic, chromium, nickel compounds, beryllium, cobalt, benzene, and various paints, dyes, tints, pesticides, and enamels.[9]

ASBESTOS HAZARDS

The Environmental Protection Agency (EPA) estimates that approximately 75 percent of the commercial buildings in use today contain asbestos in some form.[10] Asbestos was once thought to be a miracle material because of its many useful characteristics including fire resistance, heat resistance, mechanical strength, and flexibility. As a result, asbestos was widely used in commercial and industrial construction between 1900 and the mid-1970s.[11]

In the mid-1970s, medical research clearly tied asbestos to respiratory cancer, scarring of the lungs (now known as *asbestosis),* and cancer of the chest or abdominal lining *(mesothelioma).*[12] It was finally banned by the EPA in 1989.

The following quote from *Occupational Hazards* on friable asbestos shows why asbestos is still a concern even though its further use has been banned:

> When asbestos becomes friable (crumbly), it can release fibers into the air that are dangerous when inhaled. As asbestos-containing material (ACM) ages, it becomes less viable and more friable. Asbestos can be released into the air if it is disturbed during renovation or as a result of vandalism.[13]

OSHA has established an exposure threshold known as the permissible exposure limit (PEL) for asbestos. The PEL for asbestos is 0.2 fibers per cubic centimeter of air for an eight-hour time-weighted average.[14] The addresses for OSHA and other sources of further information about asbestos are given in Figure 10-4.

Removing and Containing Asbestos

When a facility is found to contain asbestos, safety and health professionals are faced with the question of whether to remove it or contain it. According to Hughes, before making this decision, the following factors should be considered:

- Is there evidence that the ACM is deteriorating? What is the potential for future deterioration?
- Is there evidence of physical damage to the ACM? What is the potential for future damage?
- Is there evidence of water damage to the ACM or spoilage? What is the potential for future damage or spoilage?[15]

Checklist of Sources of Asbestos Information

Asbestos Action Program
EPA
Mail Code TS-799
401 M Street, S.W.
Washington, D.C. 20460
202-382-3949

Cancer Information Service
National Cancer Institute
Bldg. 31, Room 10A24
9000 Rockville Pike
Bethesda, MD 20892
800-4CANCER

NIOSH Publications Office
4676 Columbia Pkwy.
Cincinnati, OH 45226
513-533-8287

OSHA Publications Office
Room N–3101
200 Constitution Avenue, N.W.
Washington, D.C. 20210
202-523-9649

Asbestos Abatement Council of AWCI
1600 Cameron Street
Alexandria, VA 22314
703-684-2924

Asbestos Information Association of
    North America
1745 Jefferson Davis Highway, Suite 509
Arlington, VA 22202
703-979-1150

The National Asbestos Council
1777 Northeast Expressway, Suite 150
Atlanta, GA 30329
404-633-2622

Figure 10–4
Sources of information about asbestos in the workplace.

Several approaches can be used for dealing with asbestos in the workplace. The most widely used are removal, enclosure, and encapsulation. These methods are explained in the following paragraphs.

Removal[16]

Asbestos removal is also known as *asbestos abatement.* The following procedures are recommended for removal of asbestos: (1) the area in question must be completely enclosed in walls of tough plastic; (2) the enclosed area must be ventilated by high-efficiency particle absolute (HEPA) filtered negative air machines (these machines work somewhat like a vacuum cleaner in eliminating asbestos particles from the enclosed area); (3) the ACM must be covered with a special liquid solution to cut down on the release of asbestos fibers; and (4) the ACM must be placed in leakproof containers for disposal.

Enclosure[17]

Enclosure of an area containing ACMs involves completely encapsulating the area in airtight walls. The following procedures are recommended for enclosing asbestos: (1) use HEPA-filtered negative air machines in conjunction with drills or any other tools that may penetrate or otherwise disturb ACMs; (2) construct the enclosing walls of impact-resistant and airtight materials; (3) post signs indicating the presence of ACMs within the enclosed area; and (4) note the enclosed area on the plans of the building.

Encapsulation[18]

Encapsulation of asbestos involves spraying the ACMs with a special sealant that binds them together, thereby preventing the release of fibers. The sealant should harden into a tough, impact-resistant skin. This approach is generally used only on acoustical plaster and similar materials.

Personal Protective Equipment for Asbestos Removal

It is important to use the proper types of personal protective clothing and respiratory devices. Clothing should be disposable and should cover all parts of the body.[19] According to Hughes, respirators used when handling asbestos should be "high-efficiency cartridge filter type (half-and-full-face types); any powered-air purifying respirator; any type C continuous-flow supplied-air, pressure-demand respirator, equipped with an auxiliary positive pressure self-contained breathing apparatus."[20]

Medical Records and Examinations[21]

It is important that employees who handle ACMs undergo periodic medical monitoring. Medical records on such employees should be kept current and maintained for at least 20 years. They should contain a complete medical history on the employee. These records must be made available on request to employees, past employees, health-care professionals, employee representatives, and OSHA personnel.

Medical examinations, conducted at least annually, should also be required for employees who handle ACMs. These examinations should include front and back chest X-rays, that are at least 7 inches by 14 inches. The examination should also test pulmonary function, including forced vital capacity and forced expiratory volume at one second.

CONFINED SPACES HAZARDS

A confined space is any area with limited means of entry and exit that is large enough for a person to fit into but is not designed for occupancy. Examples of confined spaces include vaults, vats, silos, ship compartments, train compartments, sewers, and tunnels. What makes confined spaces hazardous, beyond those factors that define the concept, is their potential to trap toxic and/or explosive vapors and gases.

In addition to the toxic and explosive hazards associated with confined spaces, there are often physical hazards. For example, tunnels often contain pipes that can trip an employee or that can leak and cause a fall. Empty liquid or gas storage vessels may contain mechanical equipment or pipes that must be carefully maneuvered around, often in the dark.

THRESHOLD LIMIT VALUES

How much exposure to a toxic substance is too much? How much is acceptable? Guidelines that answer these questions for safety and health professionals are developed and issued annually by the American Conference of Governmental Industrial Hygienists (ACGIH). The guidelines are known as threshold limit values, or TLVs. The ACGIH describes threshold limit values as follows:

> Threshold limit values refer to airborne concentrations of substances and represent conditions under which it is believed that nearly all workers may be repeatedly exposed day after day without adverse effect. Because of wide variation in individual susceptibility, however, a small percentage of workers may experience discomfort from some substances at concentrations at or below the threshold limit; a smaller percentage may be affected more seriously by aggravation of preexisting condition or by development of an occupational illness.

Threshold limits are based on the best available information from industrial experience, from experimental human and animal studies, and, when possible, from a combination of the three. The basis on which the values are established may differ from substance to substance; protection against impairment of health may be a guiding factor for some, whereas reasonable freedom from irritation, narcosis, nuisance, or other forms of stress may form the basis for others.[22]

EXPOSURE THRESHOLDS

An exposure threshold is a specified limit on the concentration of selected chemicals. Exposure to these chemicals that exceeds the threshold may be hazardous to a worker's health. Threshold recommendations are revised and updated frequently as new chemicals are introduced and as more is learned about existing chemicals. Since exposure threshold data are prone to frequent change, actual values for given chemicals are not presented in this section. Further, it is not the purpose of this section to relate any specific threshold values. Rather, this section will help prospective and practicing safety and health professionals understand the language used to express threshold values so that they can interpret threshold data, regardless of the format in which they are published.

The three most important concepts for understanding exposure thresholds are (1) time-weighted average (TWA), (2) short-term exposure limit, and (3) exposure ceiling.

- The time-weighted average (TWA) is the average concentration of a given substance to which employees may be safely exposed over an 8-hour workday or a 40-hour work week.[23]
- A short-term exposure limit is the maximum concentration of a given substance to which employees may be safely exposed for up to 15 minutes without suffering irritation, chronic or irreversible tissue change, or narcosis to a degree sufficient to increase the potential for accidental injury, impair the likelihood of self-rescue, or reduce work efficiency.[24]
- The exposure ceiling refers to the concentration level of a given substance that should not be exceeded at any point during an exposure period.[25]

Calculating a TWA

Olishifski gives the following formula for calculating the TWA for an eight-hour day:

$$TWA = \frac{CaTa + CbTb + \cdots CnTn}{8}$$

where:

Ta = time of the first exposure period during the eight-hour shift
Ca = concentration of substances in period a
Tb = another time period during the same shift
Cb = concentration during period b
Tn = nth or final time period in the eight-hour shift
Cn = concentration during period n[26]

PREVENTION AND CONTROL

Most prevention and control strategies can be placed in one of the following four categories: (1) engineering controls, (2) ventilation, (3) personal protective equipment, and (4) administrative controls.[27] Examples of strategies in each category are given in the following paragraphs.

Engineering Controls

The category of engineering controls includes such strategies as replacing a toxic material with one that is less hazardous or redesigning a process to make it less stressful or to reduce exposure to hazardous materials or conditions. Other engineering controls are isolating a hazardous process to reduce the number of people exposed to it and introducing moisture to reduce dust.[28]

Ventilation

Exhaust ventilation involves trapping and removing contaminated air. This type of ventilation is typically used with such processes as abrasive blasting, grinding, polishing, buffing, and spray painting/finishing. It is also used in conjunction with open-surface tanks. Dilution ventilation involves simultaneously removing and adding air to dilute a contaminant to acceptable levels.[29]

Personal Protection from Hazards

When the work environment cannot be made safe by any other method, personal protective equipment (PPE) is used as a last resort. PPE imposes a barrier between the worker and the hazard but does nothing to reduce or eliminate the hazard. Typical equipment includes safety goggles, face shields, gloves, boots, earmuffs, earplugs, full-body clothing, barrier creams, and respirators.[30]

Occasionally, in spite of an employee's best efforts in wearing PPE, his or her eyes or skin will be accidentally exposed to a contaminant. When this happens, it is critical to wash away or dilute the contaminant as quickly as possible. Specially designed eye wash and emergency wash stations should be readily available and accessible in any work setting where contaminants may be present.

Administrative Controls

Administrative controls involve limiting the exposure of employees to hazardous conditions using such strategies as the following: rotating schedules, required breaks, work shifts, and other schedule-oriented strategies.[31]

Additional Strategies

The type of prevention and control strategies used will depend on the evaluation of the specific hazards present in the workplace. The Society of Manufacturing Engineers recommends the following list of generic strategies that apply regardless of the setting:

- Practicing good housekeeping, including workplace cleanliness, waste disposal, adequate washing and eating facilities, healthful drinking water, and control of insects and rodents
- Using special control methods for specific hazards, such as reduction of exposure time, film badges and similar monitoring devices, and continuous sampling with preset alarms
- Setting up medical programs to detect intake of toxic materials
- Providing training and education to supplement engineering controls[32]

Self-Protection Strategies

One of the best ways to protect employees from workplace hazards is to teach them to protect themselves. Modern safety and health professionals should ensure that all employees are familiar with the following rules of self-protection:[33]

1. *Know the hazards in your workplace.* Take the time to identify all hazardous materials/conditions in your workplace and know the safe exposure levels for each.

2. *Know the possible effects of hazards in your workplace.* Typical effects of workplace hazards include respiratory damage, skin disease/irritation, injury to the reproductive system, and damage to the blood, lungs, central nervous system, eyesight, and hearing.

3. *Use personal protective equipment properly.* Proper use of personal protective equipment means choosing the right equipment, getting a proper fit, correctly cleaning and storing equipment, and inspecting equipment regularly for wear and damage.

4. *Understand and obey safety rules.* Read warning labels before using any contained substance, handle materials properly, read and obey signs, and do only authorized work.

5. *Practice good personal hygiene.* Wash thoroughly after exposure to a hazardous substance, shower after work, wash before eating, and separate potentially contaminated work clothes from others before washing them.

NIOSH GUIDELINES FOR RESPIRATORS

The respirator is one of the most important types of personal protective equipment available to individuals who work in hazardous environments. Because the performance of a respirator can mean the difference between life and death, the National Institute for Occupational Safety and Health (NIOSH) publishes strict guidelines regulating the manufacture of respirators. The standard with which manufacturers must comply is 42 C.F.R. Part 84. In addition, safety and health professionals must ensure that employees are provided respirators that meet all of the specifications set forth in 42 C.F.R. Part 84.

There are two types of respirators: air filtering and air supplying. Air-filtering respirators filter toxic particulates out of the air. To comply with 42 C.F.R. Part 84, an air-filtering respirator must protect its wearer from the most penetrating aerosol size of particle, which is 0.3 microns aerodynamic mass in median diameter. The particulate filters used in respirators are divided into three classes, each class having three levels of efficiency as shown in Figure 10–5.

Class N respirators may be used only in environments that contain no oil-based particulates. They may be used in atmospheres that contain solid or non-oil contaminates. Class R respirators may be used in atmospheres containing any contaminant. However, the filters in Class R respirators must be changed after each shift if oil-based contaminants are present. Class P respirators may be used in any atmosphere containing any particulate contaminant.

If there is any question about the viability of an air-filtering respirator in a given setting, employees should use air-supplying respirators. This type of respirator works in much the same way as an air tank for a scuba diver. Air from the atmosphere is completely blocked out, and fresh air is provided via a self-contained breathing apparatus.

Checklist of Respirator Classifications		
Class N (Not Oil Resistant)	**Class R** (Oil Resistant)	**Class P** (Oil Proof)
Efficiency 95%	95%	95%
Efficiency 99%	99%	99%
Efficiency 100%	100%	100%

Figure 10–5
Respirator classifications.

Air Safety Program Elements

Companies with facilities in which fumes, dust, gases, vapors, or other potentially harmful particulates are present should have an air safety program as part of their overall safety and health program. The program should have at least the following elements:

- Accurate hazard identification and analysis procedures to determine what types of particulates are present and in what concentration
- Standard operating procedures (in writing) for all elements of the air safety program
- Respirators that are appropriate in terms of the types of hazards present and that are 42 C.F.R. Part 84 approved
- Training including fit testing, limitations, use, and maintenance of respirators
- Standard procedures for routine procedures and storage of respirators

OSHA CHEMICAL PROCESS STANDARD

Accidental chemical releases and the explosions and fires that can subsequently result were the driving force behind the development of the OSHA Chemical Process Standard. An 1989 explosion and fire that occurred at Phillips Petroleum's Pasadena, Texas, plant killing 23 and injuring 100 people has been used as an example of the types of tragedies that OSHA is trying to prevent with this standard.[34]

The standard requires chemical producers to analyze their processes to identify potentially hazardous situations and to assess the extent of the hazard. Having done so, they must accommodate this knowledge in their emergency response plans and take action to minimize the hazards. Specific additional requirements include the following:

- Compiling process safety information
- Maintaining safe operating procedures
- Training and educating employees
- Maintaining equipment
- Conducting incident investigations
- Developing emergency response plans
- Conducting safety compliance audits[35]

OSHA CONFINED SPACE STANDARD

The OSHA standard relating to confined spaces is found in 29 C.F.R. 1910.146. This standard mandates that entry permits be required before employees are allowed to enter a potentially hazardous confined space. This means that an employee must have a written permit to enter a confined space. Before the permit is issued, a supervisor, safety/health professional, or some other designated individual should do the following:

1. *Shutdown equipment/power.* Any equipment, steam, gas, power, or water in the confined space should be shut off and locked or tagged to prevent its accidental activation.

2. *Test the atmosphere.* Test for the presence of airborne contaminants and to determine the oxygen level in the confined space. Fresh, normal air contains 20.8 percent oxygen. OSHA specifies the minimum and maximum safe levels of oxygen as 19.5 percent and 23.5 percent, respectively. Atmospheric tests indicate whether a respirator is required and, if so, what type, classification, and level.

3. *Ventilate the space.* Spaces containing airborne contaminants should be purged to remove them. Such areas should also be ventilated to keep contaminants from building up again while an employee is working in the space.

4. *Have rescue personnel stand by.* Never allow an employee to enter a confined space without having rescue personnel standing by in the immediate vicinity. These personnel should be fully trained and properly equipped. It is not uncommon for an untrained, improperly equipped employee to be injured or killed trying to rescue a colleague who gets into trouble in a confined space.

5. *Maintain communication.* An employee outside of the confined space should stay in constant communication with the employee inside. Communication can be visual, verbal, or electronic (radio, telephone) depending on the distance between the employee inside and the entry point.

6. *Use a lifeline.* A lifeline attached to a full-body harness and a block and tackle will ensure that the employee who is inside can be pulled out should he or she lose consciousness. The apparatus should be rigged so that one employee working alone can pull an unconscious employee out of the confined space.

Ventilation of Confined Spaces[36]

Before allowing employees to enter a confined space, it is important to make the space as safe as possible. One of the most effective strategies for doing so is ventilation. Because confined spaces vary in size, shape, function, and hazard potential, there must be a number of different methods for ventilating them.

Before ventilating a confined space, it should be *purged*. Purging is the process of initially clearing the space of contaminants. Once the area has been purged, ventilation can begin. Ventilation is the process of continually moving fresh air through a space. Ventilation, when properly done, will accomplish the following:

■ Dilute and replace airborne contaminants that might still be present in the confined space.

■ Ensure an adequate supply of oxygen (between 19.5 and 23.5 percent).

■ Exhaust contaminants produced by work performed in the confined space (e.g., welding, painting, and so on).

Ventilation and Local Exhaust

Providing ventilation in a confined space can maintain a comfortable temperature, it can remove odors, and it can dilute contaminants. However, never depend solely on general ventilation to remove toxic contaminants from the air. To eliminate the hazards posed by toxic contaminants such as solvent vapors and welding fumes, it is necessary to exhaust the confined space aggressively. The combination of initial purging, local exhaust, and ventilation is the ideal approach. If contaminant concentrations remain too high even with this approach, employees should wear an appropriate respirator.

Rescue Preparation[37]

The time to think about getting injured employees out of a confined space is well before they enter the space in the first place. Every year, employees are killed trying to save an injured colleague inside a confined space. In an attempt to save injured colleagues, well-meaning employees, who are neither properly trained nor adequately equipped, often fall victim to the toxic atmosphere themselves and die. This is a tragic circumstance made even more so because it is unnecessary and avoidable.

With the right amount of planning and training, employees can be quickly and effectively rescued from confined spaces. Planning should answer the following questions:

■ What types of injuries/incidents might occur in a given space?

■ What types of hazards might be present in the space?

- What precautions should be taken by rescue personnel entering the space (e.g., life-lines, hoist, respirator, and so on)?
- How much room is there to maneuver in the confined space?
- What if the victim needs first aid before he or she can be moved?

All of these questions should be answered in the organization's emergency action plan. In addition, all members of the rescue team should have received the training necessary to respond quickly, safely, and effectively. An effective response is one that is appropriate to the magnitude of the incident and is carried out safely.

OSHA STANDARDS FOR TOXIC AND HAZARDOUS MATERIALS

The OSHA standards for hazardous materials are contained in 29 C.F.R. (Subpart H). Nine of the standards apply to specific materials. Four of the standards have broader applications. The standards applying to specific materials are as follows:

Hazardous Materials (Specific Standards)

1910.101	Compressed gases
1910.102	Acetylene
1910.103	Hydrogen
1910.104	Oxygen
1910.105	Nitrous oxide
1910.108	Dip tanks
1910.109	Explosives and blasting agents. (An amendment to the OSHAct passed in 1992 now requires that manufacturers of explosives and pyrotechnics must observe the requirements of the Process Safety Management Standards in 1910.119 in addition to this standard.)
1910.110	Liquefied petroleum gases
1910.111	Anhydrous ammonia

In addition to these specific standards, Subpart H contains four standards that have broad applications. Standard 1910.106 parallels the National Fire Protection Association's NFPA 30: *Flammable and Combustible Liquids Code*. Standard 1910.107 regulates processes in which paint is applied by compressed air, electrostatic steam, or other continuous or intermittent processes. Standard 1910.119 regulates process safety management relating to 125 specific chemicals. Standard 1910.120 regulates both hazardous waste operations and spills or accidental releases.

Toxic and hazardous substances are covered in 29 C.F.R. (Subpart Z). The standards in this subpart establish permissible exposure limits (PELS) for over 450 toxic and hazardous substances. Each standard deals with a specific substance or substances. The standards contained in Subpart Z begin with 1910.1000 and run through 1910.1500.

══════ APPLICATION SCENARIOS ══════

1. Make a list of all of the airborne contaminants that you might confront in a typical day.
2. Your company plans to renovate its oldest warehouse but has learned that it contains asbestos. Contact the Asbestos Action Program in Washington, DC (see Figure 10–4) and collect all available information on what your company must do to deal with this.
3. You have been asked to send one of your employees into a confined space to work. You think there may be airborne contaminants in the space. Develop an action plan for protecting this employee.

===== ENDNOTES =====

1. Smith, R. G., and Olishifski, J. B. (Revised by C. Zuez.) *Industrial Hygiene* (Chicago: National Safety Council, 1988), p. 382.
2. Ibid., p. 17.
3. Ibid.
4. Ibid., p. 14.
5. Smith, R. G., and Olishifski, J. B. *Industrial Hygiene,* pp. 366–67.
6. Olishifski, J. B. "Overview of Industrial Hygiene," pp. 18–19.
7. Smith, R. G., and Olishifski, J. B. *Industrial Hygiene,* p. 367.
8. Ibid., p. 368.
9. Ibid., p. 369.
10. "Asbestos: Facts and Sources," *Occupational Hazards,* September 1989, p. 58.
11. Ibid.
12. Ibid.
13. Ibid.
14. Ibid.
15. Hughes, A. "Asbestos: Remove or Contain, That Is the Question," *Safety & Health,* May 1988, Vol. 137, No. 5, pp. 47–48.
16. Ibid., p. 48.
17. Ibid.
18. Ibid.
19. Ibid., p. 49.
20. Ibid.
21. Ibid.
22. American Conference of Governmental Industrial Hygienists. Preface to List of Threshold Limit Values.
23. Olishifski, "Overview of Industrial Hygiene," p. 21.
24. Ibid.
25. Ibid.
26. Ibid.
27. Olishifski, J. B. "Overview of Industrial Hygiene," p. 25.
28. Ibid.
29. Ibid., p. 26.
30. Ibid.
31. Ibid., p. 27.
32. Olishifski, J. B. "Overview of Industrial Hygiene," p. 24.
33. Florida Department of Labor and Employment Security, Division of Safety, Toxic Substances Information Center. "What You Should Know About On-the-Job Health," (Tallahassee: 1991), pp. 3–15.
34. Sheridan, P. J. "OSHA's Chemical Process Standard Sparks Controversy," *Occupational Hazards,* March 1991, p. 21.
35. Ibid., p. 22.
36. Rekus, John F. "Confined Space Ventilation," *Occupational Hazards,* March 1996, pp. 35–38.
37. Rekus, John F. "Confined Space Rescue Planning," *Occupational Hazards,* March 1996, pp. 45–48.

Radiation Hazards

The widow of a construction worker who helped build the British Nuclear Fuels (BNF) Sellafield plant was awarded $286,500 when it was determined that his death from chronic myeloid leukemia was the result of overexposure to radiation. Sellafield was constructed for the purpose of separating uranium from used fuel rods. Working at the plant for approximately nine months, the victim received a total cumulative dose of almost 52 millisieverts of radiation, which exceeded the established limit for an entire 12-month period. BNF compensated the victim's wife and the families of 20 additional workers who died from causes related to radiation.

Radiation hazards in the workplace fall into one of two categories: ionizing or nonionizing. This chapter provides prospective and practicing safety and health professionals with the information they need concerning radiation hazards in both categories.

IONIZING RADIATION: TERMS AND CONCEPTS

An ion is an electrically charged atom (or group of atoms) that became charged when a neutral atom (or group of atoms) lost or gained one or more electrons as a result of a chemical reaction. If an electron is lost during this process, a positively charged ion is produced; if an electron is gained, a negatively charged ion is produced. To ionize is to become electrically charged or to change into ions. Therefore, ionizing radiation is radiation that becomes electrically charged or changed into ions. Types of ionizing radiation are listed in Figure 11–1.

To understand the hazards associated with radiation, safety and health professionals need to understand the basic terms and concepts summarized in the following paragraphs, adapted from C.F.R. (Code of Federal Regulations) 1910.96.

Figure 11–1
Types of ionizing radiation.

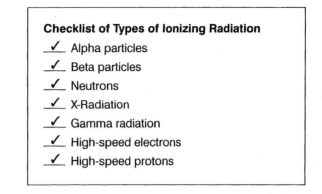

Checklist of Types of Ionizing Radiation
- ✓ Alpha particles
- ✓ Beta particles
- ✓ Neutrons
- ✓ X-Radiation
- ✓ Gamma radiation
- ✓ High-speed electrons
- ✓ High-speed protons

- Radiation consists of energetic nuclear particles and includes alpha rays, beta rays, gamma rays, X-rays, neutrons, high-speed electrons, and high-speed protons.

- Radioactive material is material that emits corpuscular or electromagnetic emanations as the result of spontaneous nuclear disintegration.

- A restricted area is any area to which access is restricted in an attempt to protect employees from exposure to radiation or radioactive materials.

- An unrestricted area is any area to which access is not controlled because there is no radioactivity hazard present.

- A dose is the amount of ionizing radiation absorbed per unit of mass by part of the body or the whole body.

- Rad is a measure of the dose of ionizing radiation absorbed by body tissues stated in terms of the amount of energy absorbed per unit of mass of tissue. One rad equals the absorption of 100 ergs per gram of tissue.

- Rem is a measure of the dose of ionizing radiation to body tissue stated in terms of its estimated biological effect relative to a dose of one roentgen (r) of X-rays.

- Air dose means that an instrument measures the air at or near the surface of the body where the highest dosage occurs to determine the level of the dose.

- Personal monitoring devices are devices worn or carried by an individual to measure radiation doses received. Widely used devices include film badges, pocket chambers, pocket dosimeters, and film rings.

- A radiation area is any accessible area in which radiation hazards exist that could deliver doses as follows: (1) within one hour, a major portion of the body could receive more than 5 millirems; or (2) within five consecutive days, a major portion of the body could receive more than 100 millirems.

- A high-radiation area is any accessible area in which radiation hazards exist that could deliver a dose in excess of 100 millirems within one hour.

EXPOSURE OF EMPLOYEES TO RADIATION

The exposure of employees to radiation must be carefully controlled and accurately monitored. Figure 11–2 shows the maximum doses for individuals in one calendar quarter. Employers are responsible for ensuring that these dosages are not exceeded.

There are exceptions to the amounts shown in Figure 11–2. According to OSHA, an employer may permit an individual in a restricted area to receive doses to the whole body greater than those shown in Figure 11–2 as long as the following conditions are met:

- During any calendar quarter, the dose to the whole body does not exceed three rems.

- The dose to the whole body, when added to the accumulated occupational dose to the whole body, shall not exceed $5(N - 18)$ rems, where N is the employee's age in years at the last birthday.

Figure 11–2
Ionizing radiation exposure limits of humans.

Checklist of Exposure Limits of Humans	
Body/Body Region	Rems Per Calendar Quarter
✓ Whole body	1.25
✓ Head and trunk	1.25
✓ Blood-forming organs	1.25
✓ Lens of eyes	1.25
✓ Gonads	1.25
✓ Hands and forearms	18.75
✓ Feet and ankles	18.75
✓ Skin of whole body	7.50

■ The employer maintains up-to-date past and current exposure records, which show that the addition of such a dose does not cause the employee to exceed the specified doses.[1]

Employers must ensure even more careful controls with individuals under 18 years of age. Such individuals may receive only doses that do not exceed 10 percent of those specified in Figure 11–2 in any calendar quarter.

OSHA is not the only agency that regulates radiation exposure. The Nuclear Regulatory Commission (NRC) is also a leading agency in this area. The NRC's regulations specify that the total internal and external dose for employees may not exceed five rems per year. This same revision established a total exposure limit of 0.6 rems over the entire course of a pregnancy for female employees. According to the NRC, the average radiation exposure of nuclear plant workers is less than 400 millirems annually.[2]

PRECAUTIONS AND PERSONAL MONITORING

Personal monitoring precautions are important for employees of companies that produce, use, release, dispose of, or store radioactive materials or any other source of ionizing radiation. Accordingly, OSHA requires the following precautions:

■ Employers must conduct comprehensive surveys to identify and evaluate radiation hazards present in the workplace from any and all sources.

■ Employers must provide appropriate personal monitoring devices such as film badges, pocket chambers, pocket dosimeters, and film rings.

■ Employers must require the use of appropriate personal monitoring devices by the following: (1) any employee who enters a restricted area where he or she is likely to receive a dose greater than 25 percent of the total limit of exposure specified for a calendar quarter; (2) any employee 18 years of age or less who enters a restricted area where he or she is likely to receive a dose greater than 5 percent of the total limit of exposure specified for a calendar quarter; and (3) any employee who enters a high-radiation area.[3]

CAUTION SIGNS AND LABELS

Caution signs and labels have always been an important part of safety and health programs. This is particularly true in companies where radiation hazards exist. The universal

Figure 11–3
Universal radiation symbol.

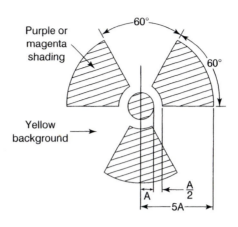

color scheme for caution signs and labels warning of radiation hazards is purple or magenta superimposed on a yellow background.

Both OSHA and the NRC require caution signs in radiation areas, high-radiation areas, airborne radiation areas, areas containing radioactive materials, and containers in which radioactive materials are stored and/or transported.[4]

Figure 11–3 shows the universal symbol for radiation. Along with the appropriate warning words, this symbol should be used on signs and labels. On containers, labels should also include the following information: (1) quantity of radioactive material, (2) kinds of radioactive materials, and (3) date on which the contents were measured.[5]

EVACUATION WARNING SIGNAL

Companies that produce, use, store, and/or transport radioactive materials are required to have a signal-generating system that can warn of the need for evacuation.[6] OSHA describes the evacuation warning signal system as follows:

> The signal shall be a mid-frequency complex sound wave amplitude modulated at a subsonic frequency. The complex sound wave in free space shall have a fundamental frequency (f_1) between 450 and 500 hertz (Hz) modulated at a subsonic rate between 4 and 5 hertz. The signal generator shall not be less than 75 decibels at every location where an individual may be present whose immediate, rapid, and complete evacuation is essential.[7]

In addition to this basic requirement, OSHA also stipulates the following:

- ■ A sufficient number of signal generators must be installed to cover all personnel who may need to be evacuated.
- ■ The signal shall be unique, unduplicated, and instantly recognizable in the plant where it is located.
- ■ The signal must be long enough in duration to ensure that all potentially affected employees are able to hear it.
- ■ The signal generator must respond automatically without the need for human activation, and it must be fitted with backup power.[8]

INSTRUCTING/INFORMING PERSONNEL

It is critical that companies involved in producing, using, storing, handling, and/or transporting radioactive materials keep employees informed concerning radiation hazards and the appropriate precautions for minimizing them. Consequently, OSHA has established specific requirements along these lines. They are summarized as follows:

- All employees must be informed of existing radiation hazards and where they exist; the extent of the hazards; and how to protect themselves from the hazards (precautions and personal protective equipment).
- All employees must be advised of any reports of radiation exposure requested by other employees.
- All employees must have ready access to C.F.R. 1910.96(i) and any related company operating procedures.[9]

These requirements apply to all companies that do not have superseding requirements (i.e., companies regulated by the Atomic Energy Commission and companies in states with their own approved state-level OSHA plans).

Instruction and information are important in all safety and health programs. They are especially important in settings in which radiation hazards exist. Employees in these settings must be knowledgeable about radiation hazards and how to minimize them. Periodic updating instruction for experienced workers is as important as initial instruction for new employees and should not be overlooked. Often, it is the overly comfortable, experienced worker who overlooks a precaution and thereby causes an accident.

STORAGE AND DISPOSAL OF RADIOACTIVE MATERIAL

Radioactive materials that are stored in restricted areas must be appropriately labeled, as described earlier in this chapter. Radioactive materials that are stored in unrestricted areas "shall be secured against unauthorized removal from the place of storage."[10] This requirement precludes the handling and transport, intentional or inadvertent, of radioactive materials by persons who are not qualified to move them safely.

A danger inherent in storing radioactive materials in unrestricted areas is that an employee, such as a maintenance worker, might unwittingly attempt to move the container and damage it in the process. This could release doses that exceed prescribed acceptable limits.

The disposal of radioactive material is also a regulated activity. There are only three acceptable ways to dispose of radioactive waste: (1) transfer to an authorized recipient;

Figure 11–4
States having agreements with the Atomic Energy Commission.

Checklist of States with Agreements with The Atomic Energy Commission

The following states have agreements with the Atomic Energy Commission to dispose of radioactive waste pursuant to 27(b) 42 U.S.C. 2021(b) of the Atomic Energy Act:

✓ Alabama	✓ Mississippi
✓ Arizona	✓ New Hampshire
✓ Arkansas	✓ New York
✓ Colorado	✓ North Carolina
✓ Florida	✓ North Dakota
✓ Georgia	✓ Oregon
✓ Idaho	✓ South Dakota
✓ Kansas	✓ Tennessee
✓ Kentucky	✓ Texas
✓ Louisiana	✓ Washington
✓ Maryland	

(2) transfer in a manner approved by the Atomic Energy Commission; or (3) transfer in a manner approved by any state that has an agreement with the Atomic Energy Commission pursuant to Section 27(b)42 U.S.C. 2021(b) of the Atomic Energy Act.[11] States having such agreements are listed in Figure 11–4.

NOTIFICATION OF INCIDENTS

A radiation-related incident must be reported if employees meet a specific set of requirements. An incident is defined by OSHA as follows:

> Exposure of the whole body of any individual to 25 rems or more of radiation; exposure of the skin of the whole body of any individual to 150 rems or more of radiation; or exposure of the feet, ankles, hands, or forearms of any individual to 375 rems or more of radiation.[12]

> The release of radioactive material in concentrations which, if averaged over a period of 24 hours, would exceed 5,000 times the limit specified. . . . [13]

If an incident meeting one of these criteria occurs, the employer must notify the proper authorities immediately. Companies regulated by the Atomic Energy Commission are to notify the commission. Companies in states that have agreements with the Atomic Energy Commission (Figure 11–4) are to notify the state designee. All other companies are to notify the U.S. assistant secretary of labor.[14] Telephone or telegraph notifications are sufficient to satisfy the immediacy requirement.

The notification requirements are eased to 24 hours in cases where whole body exposure is between 5 and 24 rems; exposure of the skin of the whole body is between 30 and 149 rems; or exposure of the feet, ankles, hands, or forearms is between 75 and 374 rems.[15]

REPORTS AND RECORDS OF OVEREXPOSURE

In addition to the immediate and 24-hour notification requirements explained in the previous section, employers are required to follow up with a written report within 30 days. Written reports are required when an employee is exposed as set forth in the previous section, or when radioactive materials are on hand in concentrations greater than specified limits. Each report should contain the following material, as applicable: extent of exposure of employees to radiation or radioactive materials; levels of radiation and concentration of radiation involved; cause of the exposure; levels of concentrations; and corrective action taken.[16]

Whenever a report is filed concerning the overexposure of an employee, the report should also be given to that employee. The following note should be placed prominently on the report or in a cover letter: "You should preserve this report for future reference."[17]

Records of the doses of radiation received by all monitored employees must be maintained and kept up to date. Records should contain cumulative doses for each monitored employee. Figure 11–5 is an example of a cumulative exposure report form of the kind that can be used to satisfy reporting requirements. Notice that the maximum quarterly dose per body or body region is indicated for that region. Such records must be shared with monitored employees at least annually.[18] A better approach is to advise monitored employees of their cumulative doses as soon as the new cumulative amount is recorded for that period.

Cumulative radiation records must be made available to former employees upon request. Upon receiving a request from a former employee, employers must provide the information requested within 30 days.[19] According to OSHA,

> Such report shall be furnished within 30 days from the time the request is made, and shall cover each calendar quarter of the individual's employment involving exposure to radiation or such lesser period as may be requested by the employee. The report shall also include the

	Rems in 1st Quarter	Rems in 2nd Quarter	Rems in 3rd Quarter	Rems in 4th Quarter	Total
Employee _____					
Dates Covered _____					
Whole body: head and trunk; blood-forming organs; lens of eyes; or gonads (1¼ rems max/quarter)					
Hands and forearms; feet and ankles (18 rems max/quarter)					
Skin of whole body (7½ rems max/quarter)					

Figure 11–5
Cumulative radiation exposure record.

results of any calculations and analysis of radioactive material deposited in the body of the employee. The report shall be in writing and contain the following statement: "You should preserve this report for future reference."[20]

NONIONIZING RADIATION

Nonionizing radiation is that radiation on the electromagnetic spectrum that has a frequency (hertz, cycles per second) of 10^{15} or less and a wavelength in meters of 3×10^{-7} or less. This encompasses visible, ultraviolet, infrared, microwave, radio, and AC power frequencies. Radiation at these frequency levels does not have sufficient energy to shatter atoms and ionize them.[21] However, such radiation can cause blisters and blindness. In

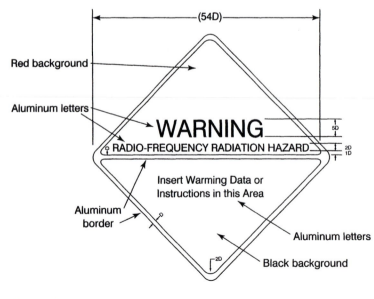

Figure 11–6
Warning symbol for radio frequency radiation.

addition, there is mounting evidence of a link between nonionizing radiation and cancer. The warning symbol for radio frequency radiation is shown in Figure 11–6.

The greatest concerns about nonionizing radiation relate to the following sources: visible radiation, ultraviolet radiation, infrared radiation, lasers, and video display terminals. The main concerns in each of these areas are explained in the following paragraphs. The following section deals specifically with electromagnetic radiation (EMR) from power lines and other sources.

- Visible radiation comes from light sources that create distortion. This can be a hazard to employees whose jobs require color perception. For example, 8 percent of the male population is red color-blind and cannot properly perceive red warning signs.[22]

- The most common source of ultraviolet radiation is the sun. Potential problems from ultraviolet radiation include sunburn, skin cancer, and cataracts. Precautionary measures include special sunglasses treated to block out ultraviolet rays and protective clothing. Other sources of ultraviolet radiation include lasers, welding arcs, and ultraviolet lamps.[23]

- Infrared radiation creates heat. Consequently, the problems associated with this kind of nonionizing radiation involve heat stress and dry skin and eyes. Primary sources of infrared radiation are high-temperature processes such as the production of glass and steel.[24]

- Lasers are being used increasingly in modern industry. The hazards of lasers consist of a thermal threat to the eyes and the threat of electrocution from power sources. In addition, the smoke created by lasers in some processes can be toxic.[25]

- Video display terminals (VDTs) are widely used in the modern workplace. They emit various kinds of nonionizing radiation. Typically, the levels are well below established standards. However, concerns persist about the long-term effects of prolonged and continual exposure to VDT-based radiation.[26] The National Institute for Occupational Safety and Health (NIOSH) undertook a study of this issue in 1984. NIOSH specifically looked into the potential impact of VDT use on pregnant users. The study found "that women who work with VDTs have no greater risk of miscarriage than those who do not."[27]

OSHA STANDARDS FOR HEALTH AND ENVIRONMENTAL CONTROLS

OSHA's standards relating to radiation hazards are contained in 29 C.F.R. 1910 (Subpart G). These standards are as follows:

Figure 11-7
Identify radiation hazards before they cause injuries.

Checklist for Identifying Radiation Hazards

✓ Are personal monitoring devices being used where appropriate?

✓ Are radiation areas clearly marked with caution signs?

✓ Are high radiation areas clearly marked with caution signs?

✓ Are airborne radiation areas clearly marked with caution signs?

✓ Are all containers in which radioactive materials are stored and/or transported clearly marked with caution signs?

✓ Is an evacuation warning signal system in place and operational?

✓ Are all containers in which radioactive materials are stored secured against unauthorized removal?

1910.94	Ventilation
1910.95	Occupational noise exposure
1910.96	Ionizing radiation
1910.97	Nonionizing radiation
1910.98	Effective dates
1910.99	Sources of standards
1910.100	Standards organizations

Figure 11-7 is a checklist that can be used to help organizations identify potential radiation hazards.

APPLICATION SCENARIOS

1. Assume you are the new safety manager of an electric power company that uses nuclear fuel to generate electricity. A quick inspection reveals that warning signs are being properly displayed, but you can find no material for informing personnel of radiation hazards and the precautions for minimizing them. When you ask your supervisor about this apparent discrepancy, he says, ""We don't have a program for keeping employees informed. That's one of the reasons we hired you. I suggest you get busy!" Develop a plan for the necessary program.

2. Continue in your role from Scenario 1. A few days later your supervisor asks the following question: "What do we need to be doing in the way of personal monitoring with our employees?" Describe what OSHA requires in this area.

3. An employee has been exposed to 1.21 Rems of radiation during the current calendar quarter. She is 17 years and 9 months old. Is this a problem?

4. Assume someone close to you is pregnant and works at a VDT all day in her job. A friend has advised her to move to another job away from VDTs until the baby is born. Her supervisor is opposed to the idea saying she is too valuable in her current position. She is concerned and comes to you for advice. What would you tell her?

ENDNOTES

1. 29 C.F.R. [Code of Federal Regulations] 1910.96(b)(2).
2. "Workers' Radiation Protection Improved by NRC Chairman Carr," *Occupational Health & Safety Letter,* December 26, 1990, Vol. 20, No. 26, p. 212.
3. 29 C.F.R. 1910.96(d).
4. 29 C.F.R. 1910.96(e).
5. 29 C.F.R. 1910.96(e)(6)(iv).
6. 29 C.F.R. 1910.96(g)(1)(2)(3).
7. 29 C.F.R. 1910.96(f)(i)(ii).
8. 29 C.F.R. 1910.96(f)(iii-vi)(2)(ii).
9. 29 C.F.R. 1910.96(i).
10. 29 C.F.R. 1910.96(j).
11. 29 C.F.R. 1910.96(k) and (i).
12. 29 C.F.R. 1910.96(l)(i).
13. 29 C.F.R. 1910.96(l)(ii).
14. 29 C.F.R. 1910.96(l)(1).
15. 29 C.F.R. 1910.96(l)(2).
16. 29 C.F.R. 1910.96(m)(1).
17. 29 C.F.R. 1910.96(m)(2).
18. 29 C.F.R. 1910.96(n)(1).

19.   29 C.F.R. 1910.96(o)(1).
20.   Ibid.
21.   "New Concerns About Nonionizing Radiation," *Occupational Hazards,* August
      1989, p. 41.
22.   Ibid., p. 42.
23.   Ibid., p. 43.
24.   Ibid.
25.   Ibid.
26.   Ibid.
27.   Castelli, J. "NIOSH Releases Results of VDT Study," *Safety & Health,* June 1991, p. 69.

CHAPTER TWELVE

Noise and Vibration Hazards

- Characteristics of Sound
- Hazard Levels and Risks
- Standards and Regulations
- Workers' Compensation and Noise Hazards
- Identifying and Assessing Hazardous Noise Conditions
- Noise Control Strategies
- Vibration Hazards

The modern industrial worksite can be a noisy place. This poses two safety- and health-related problems. First, there is the problem of distraction. Noise can distract workers and disrupt their concentration, which can lead to accidents. Second, there is the problem of hearing loss. Exposure to noise that exceeds prescribed levels can result in permanent hearing loss.

Modern safety and health professionals need to understand the hazards associated with noise and vibration, how to identify and assess these hazards, and how to prevent injuries related to them. This chapter provides the information that prospective and practicing safety and health professionals need in order to do so.

CHARACTERISTICS OF SOUND

Sound is any change in pressure that can be detected by the ear. Typically, sound is a change in air pressure. However, it can also be a change in water pressure or any other pressure-sensitive medium. Noise is unwanted sound. Consequently, the difference between noise and sound is in the perception of the person hearing it (e.g., loud rock music may be considered sound by a rock fan but noise by a shift worker trying to sleep). Olishifski describes what occurs physiologically with sound as follows:

> The generation and propagation of sound are easily visualized by means of a simple model. Consider a plate suspended in midair. When struck, the plate vibrates rapidly back and forth. As the plate travels in either direction, it compresses the air, causing a slight increase in pressure. When the plate reverses direction, it leaves a partial vacuum, or rarefaction, of the air. These alternate compressions and rarefactions cause small but repeated fluctuations in the atmospheric pressure that extend outward from the plate. When these pressure variations strike an eardrum, they cause it to vibrate in response to the slight changes in atmospheric pressure. The disturbance of the eardrum is translated into a neural sensation in the inner ear and is carried to the brain where it is interpreted as sound.[1]

Sound and vibration are very similar. Sound typically relates to a sensation that is perceived by the inner ear as hearing. Vibration, on the other hand, is inaudible and is perceived through the sense of touch. Sound can occur in any medium that has both mass and elasticity (air, water, and so on).

The unit of measurement used for discussing the level of sound and, correspondingly, what noise levels are hazardous is the decibel, or one-tenth of a *bel*. One decibel represents the smallest difference in the level of sound that can be perceived by the human ear. Figure 12–1 shows the decibel levels for various common sounds. The weakest sound that can be heard by a healthy human ear in a quiet setting is known as the threshold of hearing (10 dBA). The maximum level of sound that can be perceived without experiencing pain is known as the threshold of pain (140 dBA).

The three broad types of industrial noise are described by McDonald as follows: Wide band noise is noise that is distributed over a wide range of frequencies. Most noise from manufacturing machines is wide band noise. Narrow band noise is noise that is confined to a narrow range of frequencies. The noise produced by power tools is narrow band noise. Finally, impulse noise consists of transient pulses that can occur repetitively or nonrepetitively. The noise produced by a jackhammer is nonrepetitive impulse noise.[2]

HAZARD LEVELS AND RISKS

The fundamental hazard associated with excessive noise is hearing loss. Exposure to excessive noise levels for an extended period of time can damage the inner ear so that the ability to hear high-frequency sound is diminished or lost altogether. Additional exposure can increase the damage until even lower frequency sounds cannot be heard.[3]

In addition to hearing loss, there is evidence that excessive noise can cause other physiological problems. According to McDonald,

> Although research on the effects of noise on health is not yet complete, it appears excessive noise can cause quickened pulse, increased blood pressure, and constriction of blood vessels. These additional stresses place more burden on the heart and may lead to heart disease.[4]

A number of different factors affect the risk of hearing loss associated with exposure to excessive noise. The most important of these are

- Intensity of the noise (sound pressure level)
- Type of noise (wide band, narrow band, or impulse)
- Duration of daily exposure
- Total duration of exposure (number of years)
- Age of the individual

Figure 12–1
Selected sound levels.
*Decibels measured on the "A" weighted network (an international standardized characteristic used in sound pressure weighting).

Checklist of Selected Sound Levels	
Source	**Decibels (dBA)***
✓ Whisper	20
✓ Quiet office	50
✓ Normal conversation	60
✓ Noisy office	80
✓ Power saw	90
✓ Chain saw	90
✓ Grinding operations	100
✓ Passing truck	100
✓ Jet aircraft	150

- Coexisting hearing disease
- Nature of environment in which exposure occurs
- Distance of the individual from the source of the noise
- Position of the ears relative to the sound waves[5]

Of these various factors, the most critical are the sound level, frequency, duration, and distribution of noise. The unprotected human ear is at risk when exposed to sound levels exceeding 115 dBA. Exposure to sound levels below 80 dBA is generally considered safe. Prolonged exposure to noise levels higher than 80 dBA should be protected against through the use of appropriate personal protective devices.

To decrease the risk of hearing loss, exposure to noise should be limited to a maximum eight-hour time-weighted average of 90 dBA. McDonald provides the following general rules for dealing with noise in the workplace:

- Exposures of less than 80 dBA may be considered safe for the purpose of risk assessment.
- A level of 90 dBA should be considered the maximum limit of continuous exposure over eight-hour days without protection.
- Continuous exposure to levels of 115 dBA and higher should not be allowed.
- Impulse noise should be limited to 140 dBA per eight-hour day for continuous exposure.[6]

STANDARDS AND REGULATIONS

The primary sources of standards and regulations relating to noise hazards are OSHA and the American National Standards Institute (ANSI). OSHA regulations require the implementation of hearing conservation programs under certain conditions. OSHA's regulations should be considered as minimum standards. ANSI's standard provides a way to determine the effectiveness of hearing conservation programs such as those required by OSHA. The ANSI standard and OSHA regulations are discussed in the following sections.

ANSI Standard

In 1991, the American National Standards Institute (ANSI) published ANSI standard S12.13—1991. Entitled "Evaluation of Hearing Conservation Programs," this standard is designed to help safety and health professionals determine if hearing conservation programs work the way they are intended.[7]

Federal regulations require that employees be protected from excessive noise in the workplace. However, they provide no methodology for determining the effectiveness of hearing conservation programs. The primary reason for the development of ANSI S12.13—1991 was because hearing conservation programs were not really protecting employees but were actually only recording their steadily declining hearing ability.

OSHA Regulation

In 1983, OSHA adopted a Hearing Conservation Amendment to OSHA 29 C.F.R. 1910.95 that requires employers to implement hearing conservation programs in any work setting where employees are exposed to an eight-hour time-weighted average of 85 dBA and above.[8] Employers are required to implement hearing conservation procedures in settings where the noise level exceeds a time-weighted average of 90 dBA. They are also required to provide personal protective devices for any employee who shows evidence of hearing loss, regardless of the noise level at his or her worksite.

In addition to concerns over noise levels, the OSHA regulation also addresses the issue of duration of exposure. LaBar explains the duration aspects of the regulation as follows:

Duration is another key factor in determining the safety of workplace noise. The regulation has a 50 percent 5 dBA logarithmic tradeoff. That is, for every 5 decibel increase in the noise level, the length of exposure must be reduced by 50 percent. For example, at 90 decibels (the sound level of a lawn-mower or shop tools), the limit of "safe" exposure is 8 hours. At 95 dBA, the limit on exposure is 4 hours, and so on. For any sound that is 106 dBA and above—this would include such things as a sandblaster, rock concert, or jet engine—exposure without protection should be less than 1 hour, according to OSHA's rule.[9]

The basic requirements of OSHA's Hearing Conservation Standard are explained here:

1. *Monitoring noise levels.* Noise levels should be monitored on a regular basis. Whenever a new process is added, an existing process is altered, or new equipment is purchased, special monitoring should be undertaken immediately.

2. *Medical surveillance.* The medical surveillance component of the regulation specifies that employees who will be exposed to high noise levels be tested upon being hired and again at least annually.

3. *Noise controls.* The regulation requires that steps be taken to control noise at the source. Noise controls are required in situations where the noise level exceeds 90 dBA. Administrative controls are sufficient until noise levels exceed 100 dBA. Beyond 100 dBA, engineering controls must be used.

4. *Personal protection.* Personal protection devices are specified as the next level of protection when administrative and engineering controls do not reduce noise hazards to acceptable levels. They are to be used in addition to, rather than instead of, administrative and engineering controls.

5. *Education and training.* The regulation requires the provision of education and training to do the following: Ensure that employees understand (1) how the ear works, (2) how to interpret the results of audiometric tests, (3) how to select personal protective devices that will protect them against the types of noise hazards to which they will be exposed, and (4) how to use personal protective devices properly.[10]

WORKERS' COMPENSATION AND NOISE HAZARDS

Hearing loss claims are being covered by state workers' compensation laws. Some states have actually written hearing loss into their workers' compensation law. Others are covering claims whether hearing loss is in the law or not.

Medical professionals have established a procedure for determining if there is a causal relationship between workplace noise and hearing loss. In making determinations of such relationships, physicians consider the following factors:

1. Onset and progress of the employee's history of hearing loss
2. The employee's complete work history
3. Results of the employee's otological examination
4. Results of hearing studies that have been performed
5. Determination of whether causes of hearing loss originated outside the workplace[11]

Since approximately 15 percent of all working people are exposed to noise levels exceeding 90 dBA, hearing loss may be as significant in workers' compensation costs in the future as back injuries, carpal tunnel syndrome, and stress are now.

IDENTIFYING AND ASSESSING HAZARDOUS NOISE CONDITIONS

Identifying and assessing hazardous noise conditions in the workplace involve the following: (1) conducting periodic noise surveys, (2) conducting periodic audiometric tests, (3) record keeping, and (4) follow-up action. Each of these components is covered in the following sections.

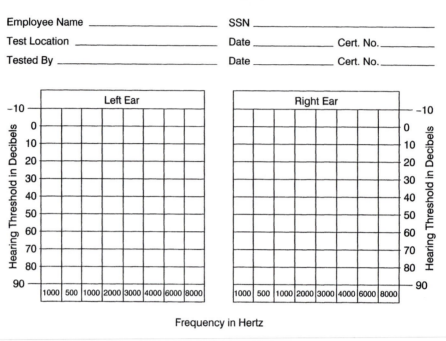

Figure 12–2
Sample audiometric test form.

Noise Surveys

Conducting noise surveys involves measuring noise levels at different locations in the workplace. The devices that are most widely used to measure noise levels are sound level meters and dosimeters. A sound level meter produces an immediate reading that represents the noise level at a specific instant in time. A dosimeter provides a time-weighted average over a period of time such as one complete work shift.[12] The dosimeter is the most widely used device because it measures total exposure, which is what OSHA and ANSI standards specify. Using a dosimeter in various work areas and attaching a personal dosimeter to one or more employees is the recommended approach to ensure dependable, accurate readings.

Audiometric Testing

Audiometric testing measures the hearing threshold of employees. Tests conducted according to ANSI S12.13—1991 can detect changes in the hearing threshold of the employee. A negative change represents hearing loss within a given frequency range.

The initial audiogram establishes a baseline hearing threshold. After that, audiometric testing should occur at least annually. Testing should not be done on an employee who has a cold, an ear infection, or who has been exposed to noise levels exceeding 80 dBA within 14 to 16 hours prior to a test. Such conditions can produce invalid results.[13]

When even small changes in an employee's hearing threshold are identified, more frequent tests should be scheduled and conducted as specified in ANSI S12.13—1991. "For those employees found to have standard threshold shift—a loss of 10 dBA or more averaged at 2,000, 3,000, and 4,000 hertz (Hz) in either ear—the employer is required to fill out an OSHA 200 form in which the loss is recorded as a worktime illness."[14]

Record Keeping

Figure 12–2 is an example of an audiometric form that can be used to record test results for individual employees. Such forms should be completed and kept on file to allow for sequential comparisons. It is also important to retain records containing a worker's employment history, including all past positions and the working conditions in those positions.

Follow-Up

Hearing loss can occur without producing any evidence of physiological damage. Therefore, it is important to follow up on even the slightest evidence of a change in an employee's hearing threshold.

Follow-up can take a number of different forms. The following would all be appropriate follow-up responses:

- Administering a retest to verify the hearing loss.
- Changing or improving the type of personal protection used.
- Conducting a new noise survey in the employee's work area to determine if engineering controls are sufficient.
- Testing other employees to determine if the hearing loss is isolated to the one employee in question or if other employees have been affected.

NOISE CONTROL STRATEGIES

Noise can be reduced by engineering and/or administrative controls applied to one or more of these components. The most desirable noise controls are those that reduce noise at the source. The second priority is to reduce noise along its path. The last resort is noise reduction at the receiver using personal protective devices. The latter approach should never be substituted for the two former approaches.

The following paragraphs explain widely used strategies for reducing workplace noise at the source, along its path, and at the receiver:

- Noise can be reduced at its source by enclosing the source, altering the acoustical design at the source, substituting equipment that produces less noise, making alterations to existing equipment, or changing the process so that less noisy equipment can be used.
- Noise can be reduced along its path by moving the source farther away from receivers and improving the acoustical design of the path so that more sound is absorbed as it travels toward receivers.
- Noise can be reduced at the receiver by enclosing the worker, using personal protective devices, and changing job schedules so that exposure time is reduced.

Some of the noise reduction strategies explained in the preceding paragraphs are engineering controls; others are administrative controls. For example, enclosing a noise source and substituting less noisy equipment are both examples of engineering controls. Changing job schedules is an example of an administrative control. Safety and health professionals should be familiar with both types of controls.

Engineering Controls

The National Safety Council describes engineering controls as "procedures other than administrative or personal protection procedures that reduce the sound level at the source or within the hearing zone of the workers."[15] What follows are commonly used engineering controls. All of these controls are designed to reduce noise at the source, along its path, or at the receiver. They focus primarily on the noise rather than the employees who are exposed to it.

Maintenance

- Replacement or adjustment of worn, loose, or unbalanced parts of machines
- Lubrication of machine parts and use of cutting oils
- Use of properly shaped and sharpened cutting tools

Substitution of Machines

- Larger, slower machines for smaller, faster ones
- Step dies for single-operation dies
- Presses for hammers
- Rotating shears for square shears
- Hydraulic presses for mechanical presses
- Belt drives for gears

Substitution of Processes

- Compression riveting for impact riveting
- Welding for riveting
- Hot working for cold working
- Pressing for rolling or forging

Reduce the Driving Force of Vibrating Surfaces by

- Reducing the forces
- Minimizing rotational speed
- Isolating

Reduce the Response of Vibrating Surfaces by

- Damping
- Additional support
- Increasing the stiffness of the material
- Increasing the mass of vibrating members
- Changing the size to change resonance frequency

Reduce the Sound Radiation from the Vibrating Surfaces by

- Reducing the radiating area
- Reducing overall size
- Perforating surfaces

Reduce the Sound Transmission Through Solids by Using

- Flexible mounting
- Flexible sections in pipe runs
- Flexible-shaft couplings
- Fabric sections in ducts
- Resilient flooring

Reduce the Sound Produced by Gas Flow by

- Using intake and exhaust mufflers

- Using fan blades designed to reduce turbulence
- Using large, low-speed fans instead of smaller, high-speed fans
- Reducing the velocity of fluid flow (air)
- Increasing the cross-section of streams
- Reducing the pressure
- Reducing the air turbulence

Reduce Noise by Reducing Its Transmission Through Air by

- Using sound-absorptive material on walls and ceiling in work areas
- Using sound barriers and sound absorption along the transmission path
- Completely enclosing individual machines
- Using baffles
- Confining high-noise machines to insulated rooms[16]

Administrative Controls

Administrative controls are controls that reduce the exposure of employees to noise rather than reducing the noise.

Administrative controls should be considered a second-level approach, with engineering controls given top priority. Smaller companies that cannot afford to reduce noise through engineering measures may use administrative controls instead. However, this approach should be avoided if at all possible.

Hearing Protection Devices

In addition to engineering and administrative controls, employees should be required to use appropriate hearing protection devices (HPDs). The following four classifications of HPDs are widely used: enclosures, earplugs, superaural caps, and earmuffs.

Enclosures are devices that completely encompass the employee's head, much like the helmets worn by jet pilots. Earplugs (also known as *aurals)* are devices that fit into the ear canal. Custom-molded earplugs are designed and molded for the individual employee. Premolded earplugs are generic in nature, are usually made of a soft rubber or plastic substance, and can be reused. Formable earplugs can be used by anyone. They are designed to be formed individually to a person's ears, used once, and then disposed.

Superaural caps fit over the external edge of the ear canal and are held in place by a headband. Earmuffs, also known as *circumaurals,* cover the entire ear with a cushioned cup that is attached to a headband. Earplugs and earmuffs are able to reduce noise by 20 to 30 decibels. By combining earplugs and earmuffs, an additional three to five decibels of blockage can be gained.

The effectiveness of HPDs can be enhanced through the use of technologies that reduce noise levels. These Active Noise Reduction (ANR) technologies reduce noise by manipulating sound and signal waves. Such waves are manipulated by creating an electronic mirror image of sound waves that tends to cancel out the unwanted noise in the same way that negative numbers cancel out positive numbers in a mathematical equation. Using ANR in conjunction with enclosure devices or earmuffs can be an especially effective strategy.

Traditional, or passive, HPDs can distort or muffle sounds at certain frequencies, particularly high-pitched sounds. Flat-attenuation HPDs solve this problem by using electronic devices to block all sound frequencies equally. This eliminates, or at least reduces, the distortion and muffling problems. Flat-attenuation HPDs are especially helpful for employees in settings where high-pitched sound is present and should be heard and for employees who have already begun to lose their ability to hear such sounds. The ability to hear high-pitched sounds is significant because warning signals and human voices can be high pitched.

A benefit of ANR technologies is optimization. The amount of noise protection can be adjusted so that employees can hear as much as they should, but not too much. Too much noise can cause employees to suffer hearing loss. Too little noise can mean that they may not hear warning signals.

VIBRATION HAZARDS

Vibration hazards are closely associated with noise hazards because tools that produce vibration typically also produce excessive levels of noise. The strategies for protecting employees against the noise associated with vibrating tools are the same as those presented so far in this chapter. This section focuses on the other safety and health hazards associated with vibration.

The types of injuries associated with vibration depend on its source. For example, workers who operate heavy equipment often experience vibration over the whole body. This can lead to problems ranging from motion sickness to spinal injury. However, the most common vibration-related problem is known as hand-arm vibration syndrome, or HAV. Eastman describes HAV as follows:

> The condition, a form of Reynaud's Syndrome, strikes an alarming number of workers who use vibrating power tools day in and day out as part of their jobs. For HAV sufferers, . . . the sensations in their hands are more than just minor, temporary discomforts. They are symptoms of the potentially irreversible damage their nerves and blood vessels have suffered. As the condition progresses, it takes less and less exposure to vibration or cold to trigger the symptoms, and the symptoms themselves become more severe and crippling.[17]

Environmental conditions and worker habits can exacerbate the problems associated with vibration. For example, working with vibrating tools in a cold environment is more dangerous than working with the same tools in a warm environment. Gripping a vibrating tool tightly will lead to problems sooner than using a loose grip. Smoking and excessive noise also increase the potential for HAV and other vibration-related injuries. What all of these conditions/habits have in common is that they constrict blood vessels which in turn restricts blood flow to the affected part of the body.[18]

Injury Prevention Strategies

Modern safety and health professionals should know how to prevent vibration-related injuries. Prevention is especially important with HAV because the disease is thought to be irreversible. This does not mean that HAV cannot be treated. It can, but the treatments developed to date only reduce the symptoms. They do not cure the disease.

Following are prevention strategies that can be used by safety and health professionals in any company regardless of its size.

Purchase Low-Vibration Tools

Interest in producing low-vibration tools is relatively new but growing. In the past, only a limited number of manufacturers produced low-vibration tools. However, a lawsuit filed against three prominent tool manufacturers generated a higher level of interest in producing low-vibration tools.

According to Eastman, the suit was filed on behalf of 300 employees of General Dynamic's Electric Boat Shipyard in Connecticut, most of whom now suffer from HAV. The case claimed that three predominant tool manufacturers failed to (1) warn users of their tools of the potential for vibration-related injury; and (2) produce low-vibration tools even though the technology to do so has been available for many years.[19] As a

---

Checklist for Identifying Vibration Hazards

✓ What are the sound level meter readings taken at different times and locations?

✓ What are the dosimeter readings for each shift of work?

✓ Are audiometer tests being conducted at appropriate intervals?

✓ What are the results of audiometric tests?

✓ Are records of audiometric tests being properly maintained?

✓ What is being done to reduce noise at the source?

✓ What is being done to reduce noise along its path?

✓ Are workers wearing personal protective devices where appropriate?

✓ Are low-vibration tools being used wherever possible?

✓ Are employees who use vibrating tools or equipment doing the following as appropriate?

- Wearing properly fitting thick gloves?
- Taking periodic breaks?
- Using a loose grip on vibrating tools?
- Keeping warm?
- Using vibration-absorbing floor mats and seat covers as appropriate?

Figure 12–3
Detect vibration hazards and potential hazards before they cause injuries.

result of this lawsuit and the potential for others like it, low-vibration tools have become more commonplace.

Limit Employee Exposure

Although a correlation between cumulative exposure to vibration and the onset of HAV has not been scientifically quantified, there is strong suspicion in the safety and health community that such a link exists. For example, NIOSH recommends that companies limit the exposure of their employees to no more than four hours per day, two days per week.[20] Until the correlation between cumulative exposure and HAV has been quantified, safety and health professionals are well advised to apply the NIOSH recommendation.

Change Employee Work Habits

Employees can play a key role in protecting themselves if they know how. Safety and health professionals should teach employees who use vibration-producing tools the work habits that will protect them from HAV and other injuries. These work habits include the following: (1) wearing properly fitting thick gloves that can partially absorb vibration; (2) taking periodic breaks (at least ten minutes every hour); (3) using a loose grip on the tool and holding it away from the body; (4) keeping tools properly maintained (i.e., replacing vibration-absorbing pads regularly); (5) keeping warm; and (6) using vibration-absorbing floor mats and seat covers as appropriate.[21]

Modern safety and health professionals should also encourage higher management to require careful screening of applicants for jobs involving the use of vibration-producing tools and equipment. Applicants who smoke or have other conditions that constrict blood vessels should be guided away from jobs that involve excessive vibration.

Figure 12–3 is a checklist that can be used to help organizations identify potential vibration hazards.

APPLICATION SCENARIOS

1. A friend knows that you are the safety manager for ABC Company. He is concerned that noise hazards at his job could be causing hearing loss in employees. He wants to know what factors should be considered in working with employees on an individual basis to determine hazard risks. What factors would you recommend that he consider?

2. As the safety director for XYZ Company, you are responsible for developing a hearing conservation program that will be adopted company-wide. Develop a plan for such a program.

3. As the safety engineer for Precision Manufacturing, Inc., you are responsible for developing a list of engineering controls for reducing the noise hazard level on the company's shop floor. The shop area contains hydraulic presses and punches, lathes, milling machines, metal saws, and drills. Develop the list of engineering controls that you will recommend.

4. Develop a vibration injury-prevention program for a company that is experiencing a high rate of employees experiencing hand-arm vibration syndrome.

ENDNOTES

1. Olishifski, J. B. (Revised by J. J. Standard.) *Fundamentals of Industrial Hygiene* (Chicago: National Safety Council, 1988), pp. 165–66.
2. McDonald, O. F. "Noise: How Much Is Too Much," *Safety & Health*, November 1987, Vol. 136, No. 5, p. 37.
3. Ibid.
4. Ibid.
5. Olishifski, *Fundamentals of Industrial Hygiene*, p. 171.
6. McDonald, "Noise: How Much Is Too Much," p. 38.
7. Ruck, D. "ANSI Dams the Hearing Loss Tide," *Safety & Health*, July 1991, Vol. 144, No. 1, pp. 40–43.
8. LaBar, G. "Sound Policies for Protecting Workers' Hearing," *Occupational Hazards*, July 1989, p. 46.
9. Ibid.
10. Ibid., p. 48.
11. Olishifski, *Fundamentals of Industrial Hygiene*, pp. 163–64.
12. Breisch, S. L. "Hear Today and Hear Tomorrow," *Safety & Health*, June 1989, Vol. 139, No. 6, p. 40.
13. Ibid., p. 42.
14. Ibid.
15. Olishifski, *Fundamentals of Industrial Hygiene*, pp. 178–79.
16. Ibid., p. 179.
17. Ibid.
18. Ibid., pp. 32–33.
19. Ibid., p. 34.
20. Ibid., p. 33.
21. Ibid., p. 35.

Automation and Technology Hazards

IMPACT OF AUTOMATION ON THE WORKPLACE

The advent of automation in the workplace was the next logical step on a continuum of developments intended to enhance productivity, quality, and competitiveness. This continuum began when humans first developed simple tools to assist them in doing work. This was the age of hand tools and manual work. It was eventually superseded by the age of mechanization during the Industrial Revolution. During the age of mechanization, machines were developed to do work previously done by humans using hand tools. The 1960s saw the beginnings of broad-based efforts at automating mechanical processes and systems.

These early attempts at automation resulted in islands of automation, or individual automated systems lacking electronic communication with other related systems. Examples of islands of automation are a stand-alone computer numerical control milling machine or a personal computer-based word processing system, neither of which is connected to other related systems. Local area networks (LANs) for integrating personal computers are an example of integration in the office. Computer-integrated manufacturing is an example in the factory.

These developments are having an impact on the workplace. According to Ebukuro, automation and integration are having the following effects on workers:

- Reducing the amount of physical labor workers must perform
- Increasing the amount of mental work required
- Polarizing work into mental jobs and labor-intensive jobs
- Increasing the stress levels of managers
- Decreasing the need for traditional blue-collar workers
- Decreasing the feelings of loyalty that workers feel toward employers
- Increasing workers' feelings of powerlessness and helplessness[1]

These various effects of automation are resulting in a marked increase in the amount of stress experienced by workers. Two factors in particular lead to increased levels of stress: rapid, continual change and an accompanying feeling of helplessness. With automation, the rate of change has increased. As a result, workers must continually learn and relearn their jobs with little or no relief. In addition, automated machines do more and more of the work that used to be done by humans. This can leave workers feeling as if they might be replaced by a machine and powerless to do anything about it.

Work stress is a complex concept involving physiological, psychological, and social factors. People become stressed when there is an imbalance between the demands placed on them and their ability to respond.[2] Automation appears to be increasing the instances in which such an imbalance occurs. This chapter focuses on the safety and health concerns associated with computers, robots, and automation and appropriate measures for dealing with these concerns.

VDTs IN OFFICES AND FACTORIES

A safety and health concern brought about by the advent of computers has to do with the impact of video display terminals (VDTs). Does prolonged use of VDTs cause safety and health problems? Are pregnant women who work at VDTs more likely to miscarry? Are such problems as eye fatigue, muscle stiffness, and mental fatigue caused by VDT use?

The National Institute for Occupational Safety and Health (NIOSH) published a study showing that "women who work with VDTs have no greater risk of miscarriage than those who do not."[3] According to Castelli,

> NIOSH researchers studied telephone operators in eight southeastern states. One group did not use computer terminals; the other group, directory-assistance operators, used them seven hours a day. The researchers interviewed 2,430 operators and gathered information on 882 pregnancies. . . . About half of the pregnant workers used VDTs.[4]

The miscarriage rate among VDT users was actually lower than that for nonusers (14.8 percent for VDT users and 15.9 percent among nonusers). Both figures are close to the national average of 15 percent.

Not everyone agreed with the findings of the NIOSH study, however. The Service Employees' International Union (SEIU) challenged the findings as inconclusive because the effects of stress and VDT use on pregnancy were not considered. SEIU claimed that the study did not give VDTs a "clean bill of health" and that "employers should likewise not conclude from this study that they have no obligation to develop safety guidelines and better designed equipment."[5]

Regardless of which side of the miscarriage issue one believes, there is ample evidence that such problems as eye fatigue, blurred vision, eye strain, and nervousness are associated with VDT use.

The eyestrain caused by prolonged VDT use poses safety and health problems from two different perspectives. First, the visual health of the operator experiencing the strain is impaired. Second, there is an increased likelihood of accidents caused by impaired work performance and increased psychological stress.

According to Kurimoto and associates of the Department of Ophthalmology at Japan University of Occupational and Environmental Health, the eye functions most noticeably affected by VDT use are accommodation, convergence, and lacrimation.[6] Accommodation is the ability of the eye to become adjusted after viewing the VDT so as to be able to focus on other objects, particularly objects at a distance. Convergence is the coordinated turning of the eyes inward so as to focus on a nearby point or object. Lacrimation is the process of excreting tears. The Kurimoto research confirms in nonmedical terms that prolonged VDT use can render operators unable to focus on either distant or near objects. It can also impair the tearing function, leading to dry eyes.[7]

Kurimoto recommends the following strategies for reducing the physiological and psychological problems associated with VDT use:[8]

- Faster computer response time. This is a matter of upgrading the computer's processing capability or replacing it with one that has more processing power.

- More frequent breaks from VDT use or a work rotation schedule that allows users to intersperse non-VDT work in their daily routine.

- Work design that recognizes and accommodates the need to break up continual VDT use.

- Arranging the keyboard properly so it is located in front of the user, not to the side. Body posture and the angle formed by the arms are critical factors.

- Adjusting the height of the desk. Taller employees often have trouble working at average height desks.

- Adjusting the tilt of the keyboard. The rear portion of the keyboard should be lower than the front.

- Encouraging employees to use a soft touch on the keyboard and when clicking a mouse. A hard touch increases the likelihood of injury.

- Encouraging employees to avoid wrist resting. Resting the wrist on any type of edge can increase pressure on the wrist.

- Placing the mouse within easy reach. Extending the arm to its full reach increases the likelihood of injury.

- Removing dust from the mouse ball cavity. Dust can collect, making it difficult to move the mouse. Blowing out accumulated dust once a week will keep the mouse easy to manipulate.

- Locating the VDT at a proper height and distance. The height of the VDT should be such that the top line on the screen is slightly below eye level. The optimum distance between the VDT and user will vary from employee to employee, but will usually be between 16 and 32 inches.

- Minimizing glare. Glare from a VDT can cause employees to adopt harmful postures. Glare can be minimized by changing the location of the VDT, using a screen hood, and closing or adjusting blinds and shades.

- Reducing lighting levels. Vision strain can be eliminated by reducing the lighting level in the area immediately around the VDT.

- Dusting the VDT screen. VDT screens are magnets to dust. Built-up dust can make the screen difficult to read, contributing to eyestrain.

- Eliminating telephone cradling. Cradling a telephone receiver between an uplifted shoulder and the neck while typing can cause a painful disorder called cervical radiculopathy (compression of the cervical vertebrae in the neck). Employees who need to talk on the telephone while typing should wear a headphone.

HUMAN/ROBOT INTERACTION

Every new tool developed to enhance the ability of humans to work efficiently and effectively has brought with it a new safety and health hazard. This is particularly the case with industrial robots. What makes robots more potentially dangerous than other machines can be summarized as follows: (1) their ability to acquire intelligence through programming; (2) their flexibility and range of motion; (3) their speed of movement; and (4) their power.

The often-discussed peopleless factory is still far in the future. However, robots are so widely used now that they are no longer the oddity they once were. Consequently, there is plenty of human/robot interaction in modern industry. According to Yamashita,

"At ordinary factories . . . human workers and robots coexist, creating such problems as cooperation and competition between man and machines and safety."[9]

How does human/robot interaction differ from human interaction with other machines? This is an important question for safety and health professionals. According to Lena Martensson of the Royal Institute of Technology in Stockholm, Sweden, the modern factory has or is moving toward having the following characteristics:

- Workers will supervise machine systems rather than interact with individual pieces of production equipment.
- Workers will communicate with machines via video display terminals on which complex information processed by a computer will be displayed.
- Workers will be supported by expert systems for fault identification, diagnosis, and repair.[10]

Robots and other intelligent computer-controlled machines will play an increasingly important role in modern industry. As this happens, safety and health professionals must be concerned about the new workplace hazards that will be created.

SAFETY AND HEALTH PROBLEMS ASSOCIATED WITH ROBOTS

Robots are being used in industry for such applications as arc welding, spot welding, spray painting, material handling and assembly, and loading/unloading of machines. According to the National Safety Council the principal hazards associated with robots are as follows:

- Being struck by a moving robot while inside the work envelope. The work envelope of a robot is the total area within which the moving parts of the robot actually move. Figure 13–1 is an example of a robot's work envelope.
- Being trapped between a moving part of a robot and another machine, object, or surface.
- Being struck by a workpiece, tool, or other object dropped or ejected by a robot.[11]

Until a worker enters the work envelope of a robot, there is little probability of an accident. However, any time a worker enters a functioning robot's work envelope, the probability becomes very high. The only logical reason for a worker to enter the work envelope of an engaged robot is to teach it a new motion.

Minimizing the Safety and Health Problems of Robots

If human workers never had to enter a work envelope, the safety and health problems associated with robots would be minimal. However, workers must occasionally do so. Therefore safety and health professionals must be concerned with ensuring safe human/robot interaction.

The National Safety Council recommends several strategies for minimizing the hazards associated with robots. The general strategies are summarized as follows:

- Ensure a glare-free, well-lighted robot site. The recommended light intensity is 50–100 foot-candles.
- Keep the floors in and around the robot site carefully maintained, clean, and free of obstructions so that workers do not trip or slip into the work envelope.
- Keep the robot site free of associated hazards such as blinding light from welding machines or vapors from a paint booth.
- Equip electrical and pneumatic components of the robot with fixed covers and guards.
- Clear the work envelope of all nonessential objects and make sure all safeguards are in place before starting the robot.

Figure 13–1
A robot's work envelope.
Courtesy of Unimation, Inc.

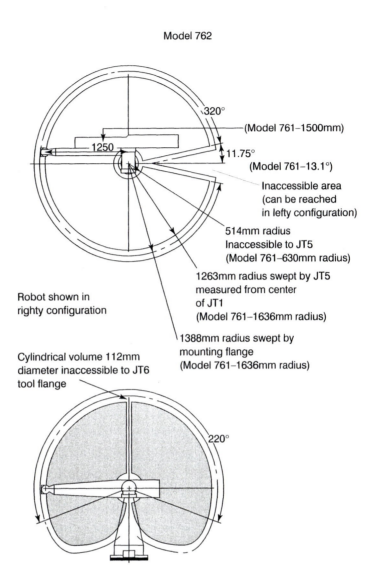

Model 762

320°
(Model 761–1500mm)

1250

11.75°
(Model 761–13.1°)

Inaccessible area
(can be reached
in lefty configuration)

514mm radius
Inaccessible to JT5
(Model 761–630mm radius)

1263mm radius swept by JT5
measured from center
of JT1
(Model 761–1636mm radius)

Robot shown in
righty configuration

1388mm radius swept by
mounting flange
(Model 761–1636mm radius)

Cylindrical volume 112mm
diameter inaccessible to JT6
tool flange

220°

- Apply lockout and proper test procedures before entering the work envelope.
- Remove and account for all tools and equipment used to maintain the robot before starting it.[12]

SAFETY AND HEALTH IN OFFICE AUTOMATION

According to Nishiyama, the objectives of office automation are increased efficiency, personnel reductions, economy of personnel expenditures, improved service to customers, improved planning and estimating, increased processing speed, and improved working conditions.[13] These goals are being achieved with varying degrees of success. However, automation also introduced a new set of safety and health problems into the office environment.

Morooka and Yamoda of Japan's Tokai University identified the following problems associated with office automation: eye fatigue, seeing double images and complementary colors, headache, yawny feelings, unwillingness to talk, shoulder fatigue, neck fatigue, dryness in the throat, sleepy feelings, and whole body tiredness.[14]

As such problems began to be associated with office automation, interest grew in establishing workplace and technology standards to minimize them. Benjamin C. Amick III of Congress' Office of Technology Assessment described this situation as follows:

> The public policy issues developing as a result of office automation range from issues of labor management relations to office, work-station, and human-computer interface standards. The current scientific research suffers from a lack of integration and clear definition of what is causing what. . . . Clearly, the current upswing in the purchasing of office automation equipment provides a unique opportunity to effect changes in the quality of worklife on a national basis. The question facing policy-makers is how best to create the policy to maintain the current level of creativity and innovation in the marketplace while not sacrificing the quality of worklife.[15]

States that have introduced legislation establishing standards relating to office automation are Oregon, Washington, California, Florida, Colorado, Missouri, Iowa, Minnesota, Wisconsin, Illinois, Indiana, Ohio, Pennsylvania, West Virginia, Maryland, New Jersey, Connecticut, Rhode Island, Massachusetts, New Hampshire, and Maine. New Mexico established standards by executive order.

Most legislation dealing with office automation concerns standards for VDT interaction. For example, legislation introduced in Maryland requires that VDT users (1) have an eye examination every year; (2) have an adjustable chair with adjustable backrest height and tension; and (3) take a 15-minute break from the VDT every hour. Legislation introduced in other states is similar to the Maryland proposal.[16]

Public policy debate is increasing in the United States over the safety and health concerns inherent in office automation. Unfortunately, the debate has outpaced the research on this important issue. Consequently, the probability is high that policies being adopted are based on insufficient information. Regarding research that needs to be conducted, Amick recommends the following:

- Testing of the biological plausibility of health hazards from VDT work
- Ergonomic intervention research to determine the contribution of workstation design and office design to the worker's health
- Examination of the interaction of physical and psychosocial stressors in the high-technology workplace
- Development of prospective case-control studies to determine the temporal relationships
- Establishment of a national high-technology surveillance system for worker safety and health
- Development of organizational and job-intervention programs
- Multidisciplinary studies examining the relative contribution of various working conditions to the health and well being of the office worker
- Programs that bring system designers and/or building designers into the total automation process[17]

The Japanese Association of Industrial Health developed the following set of principles upon which future safety and health measures relating to office automation should be based:

- More attention should be focused on bringing employers and employees together for the purpose of improving working conditions in automated offices.
- Since VDT use is becoming so common in so many different occupations, it should no longer be considered specialized. Consequently, jobs should be designed to accommodate VDT use.
- The amount of time spent doing VDT work exclusively should be kept short, and employees should be able to perform the work in their own way.
- Working conditions should be established that prevent safety and health problems so that management is acting instead of reacting.

■ Special emphasis should be placed on education and training as the best way to prevent adverse effects from office automation.

■ VDT work should not be done part-time at home or contracted out since working conditions cannot be properly controlled or supervised under these circumstances.[18]

TECHNOLOGICAL ALIENATION IN THE AUTOMATED WORKPLACE

As technology has become more widely used in the workplace, particularly automated technology, some workers have come to resent its impact on their lives. This concept is known as technological alienation. According to Gary Benson of the University of Wyoming, this concept has several meanings, all of which encompass one or more of the following:

■ Powerlessness is the feeling that workers have when they are not able to control the work environment. Powerless workers may feel that they are less important than the technology with which they work and that they are expendable.

■ Meaninglessness is the feeling that workers get when their jobs become so specialized and so technology-dependent that they cannot see the meaning in their work as it relates to the finished product or service.

■ Normlessness is the phenomenon in which people working in a highly automated environment can become estranged from society. Normless people lose sight of societies, norms, rules, and mores.[19]

Benson investigated what he considers to be the most devastating form of technological alienation—mindlessness.[20] Mindlessness is the result of the process of "dumbing down" the workplace. This is a concept that accompanied automation. In the past, machines have been used to do physical work previously done by human workers. With the advent of computers, robots, and automation, machines began doing mental work. According to Benson,

> The net result is jobs and work environments where people do not have to use their minds or think to do their work—an environment where computers, robots, and other forms of high technology do the thinking.[21]

Mindlessness on the job should be of interest to safety and health professionals because of the other problems it can create. According to Benson, these problems can include an increase in alcoholism, drug abuse, employee theft, work-related accidents, absenteeism, sick leave abuse, turnover rates, and employee personal problems. Mindlessness can also lead to a decrease in job performance, productivity, and work quality.[22]

MINIMIZING THE PROBLEMS OF AUTOMATION

The infusion of technology into the workplace has presented safety and health professionals with an entirely new set of challenges. Among the most pressing of these is the need to identify and minimize the new safety and health problems specifically associated with automation. Behavioral scientist A. B. Cherns developed a sociotechnical system theory for doing so that consists of the following components: variance control, boundary location, work group organization, management support, design process, and quality of work life.[23]

Although the sociotechnical system theory was developed in 1977, it has even more relevance now than it did then. According to Yoshio Hayashi of Japan's Keio University,

> The safety and health of workers in this high technology age cannot be discussed within the conventional framework of one worker assigned to one machine. . . . The socio-technical sys-

tem may be roughly understood if the *man* and *machine* in the man-machine system are replaced by *socio* and *technical* respectively. It refers to a system composed of a work group and high technology.[24]

The various components in the sociotechnical system theory explain what must happen if humans and technological systems are going to work together harmoniously and safely. Safety and health professionals can apply the theory as they work to minimize the potential problems associated with automation in the modern workplace. These components can be summarized as follows:

- Variance control involves controlling the unexpected events that can be introduced by new technologies. For example, a runaway, out-of-control industrial robot introduces unexpected safety hazards at variance with the expectations of workers and management. Variance control involves bringing the situation under control and establishing appropriate preventive measures for the future.

- The concept of boundary location involves the classification of work. What specific tasks are included in an employee's job description? Does a robot technician just operate the robot, or is she also required to teach and maintain the robot? The accident prevention measures learned by an employee should cover all tasks in his or her job description.

- The concept of work group organization involves identifying the tasks that a work group is to perform and how these tasks are to be performed. The key is to make sure that all work group members have the training needed to accomplish effectively and safely all tasks assigned to them.

- Management support is perhaps the most important of the components of the sociotechnical system theory. It states that, in the age of high technology, managers must be willing to accept occasional temporary declines in productivity without resorting to shortcuts or improvement efforts that might be unsafe or unhealthy. Management must be willing to emphasize safety in spite of temporary declines in productivity.

- The design process component refers to the ability of an organization to design itself in ways that promote productivity, quality, competitiveness, safety, and health. It also involves the ability to continually redesign as technological advances and other circumstances dictate.

- Quality of work life involves determining ways to promote the morale and best interests of workers. The key is to ensure that technology extends the abilities of humans and that technological systems are *human-centered*. In other words, it is important to ensure that people control systems rather than vice versa.[25]

If the sociotechnical system theory is fully applied, the safety and health hazards of the automated workplace can be minimized. Safety and health professionals can play a key role in making sure the theory is applied. To play such a role, these professionals must be technicians, diplomats, trainers, and lobbyists. They must work with the technical aspects of variance control, boundary location, work group organization, and the design process. They must be diplomats in working with supervisors and employees in promoting adherence to safe work practices. They must be trainers in order to ensure that all employees know how to apply safe work practices and appropriate accident prevention techniques. Finally, they must be lobbyists as they continually interact with management to establish and maintain management support for safety, health, and quality of life issues.

Safety Measures for Automated Systems

The sociotechnical system theory discussed in the previous section is broad and conceptual in nature. Modern safety and health professionals also need to know specific measures that can be taken to minimize the hazards associated with robots and other automated systems. Minoru Goto of the Nissan Motor Company's safety department

---

Checklist for Identifying Automation Hazards

✓ Are workers who use VDTs as their primary work tool employing methods to prevent eye strain?

✓ Are all robot sites well-lighted and marked off?

✓ Are the floors around robot sites clean and properly maintained so that workers won't slip and fall into the robot's work envelope?

✓ Are all electronic and pneumatic components of all robots equipped with guards and covers?

✓ Are the work envelopes of all robots equipped with guards and covers?

✓ Are automatic shut-off systems in place for all robots and other automated systems?

✓ Are lockout systems used before workers enter a robot's work envelope?

✓ Are safety fences erected around all automated systems?

✓ Are the controllers for all automated systems located outside of work envelopes?

Figure 13–2
Never overlook the potential hazards of automation.

developed specific safety measures in the categories of technological systems, auxiliary equipment, and training.[26]

Examples of safety measures that can be used at the technological systems level include the following:

■ Construction of a safety fence around the system that defines the work envelope of the system

■ Control of the speed of movement of system components when working inside the work fence

■ Installation of an emergency stop device colored red and placed in an easily accessible location

■ Location of the control panel for the system outside of the safety fence

■ Establishment of automatic shutdown switches that activate any time a system component goes beyond its predetermined operational range[27]

Safety measures relating to training include training system operators to work safely within the work envelope and to work together as a team when interacting with the system. Maintenance workers should be trained on the technical aspects of maintaining all machines and equipment that make up the system. This is important because the safety level of the system is the sum of the safety levels of its individual components. A system with four properly operating components and just one faulty component is an unsafe system.[28] Figure 13–2 is a checklist for identifying automation hazards.

APPLICATION SCENARIOS

1. Defend or refute the following statements: "Automation is one of the best developments in the history of the workplace. It has been positive. There is no down side."

2. Explain how you would use work design to reduce the hazards associated with continuous VDT use.

3. Assume that your new employer is having a disturbing number of accidents involving robots. Develop a plan for preventing robot-related accidents.

4. Develop a generic plan that any organization could use to minimize the problems of automation.

=== ENDNOTES ===

1. Ebukuro, R. "Alleviation of the Impact of Microelectronics on Labour," in *Occupational Health and Safety in Automation and Robotics,* ed. K. Noro (Chicago: National Safety Council, 1987), p. 11.
2. Ibid., p. 17.
3. Castelli, J. "NIOSH Releases Results of VDT Study," *Safety & Health,* June 1991, Vol. 143, No. 6, p. 69.
4. Ibid.
5. Business Publisher, Inc. *Occupational Health & Safety Letter,* March 20, 1991, Vol. 21, No. 6, p. 49.
6. Kurimoto, S., Tsuneto, I., Kageyu, N., Yamamoto, S., and Komatsubara, A. "Eye Strain in VDT Work from the Standpoint of Ergophthalmology," in *Occupational Health and Safety in Automation and Robotics,* ed. K. Noro (Chicago: National Safety Council, 1987), p. 112.
7. Ibid., pp. 112–33.
8. Ibid., p. 133.
9. Yamashita, T. "The Interaction Between Man and Robot in High Technology Industries," in *Occupational Health and Safety in Automation and Robotics,* ed. K. Noro (Chicago: National Safety Council, 1987), p. 140.
10. Martensson, L. "Interaction Between Man and Robots with Some Emphasis on 'Intelligent' Robots," in *Occupational Health and Safety in Automation and Robotics,* ed. K. Noro (Chicago: National Safety Council, 1987), p. 144.
11. National Safety Council. *Robots,* Data Sheet 1–717–85 (Chicago: National Safety Council, 1991), p. 1.
12. Ibid., pp. 2–3.
13. Nishiyama, K. "Introduction and Spread of VDT Work and Its Occupational Health Problems in Japan," in *Occupational Health and Safety in Automation and Robotics,* ed. K. Noro (Chicago: National Safety Council, 1987), p. 251.
14. Morooka, K., and Yamoda, S. "Multivariate Analysis of Fatigue on VDT Work," in *Occupational Health and Safety in Automation and Robotics,* ed. K. Noro (Chicago: National Safety Council, 1987), p. 236.
15. Amick, B, III. "The Impacts of Office Automation on the Quality of Worklife: Considerations for United States Policy," in *Occupational Health and Safety in Automation and Robotics,* ed. K. Noro (Chicago: National Safety Council, 1987), p. 232.
16. Ibid., p. 229.
17. Ibid., p. 223.
18. Ibid., pp. 261–62.
19. Benson, G. "Mindlessness: A New Dimension of Technological Alienation—Implications for the Man-Machine Interface in High Technology Work Environments," in *Occupational Health and Safety in Automation and Robotics,* ed. K. Noro (Chicago: National Safety Council, 1987), pp. 326–27.
20. Ibid., p. 328.
21. Ibid.
22. Ibid., p. 332.
23. Cherns, A. B. "Can Behavioral Science Help Design Organizations?" *Organizational Dynamics,* Spring 1977, pp. 44–64.
24. Hayashi, Y. "Measures for Improving the Occupational Health and Safety of People Working with VDTs or Robots—Small-Group Activities and Safety and Health Education," in *Occupational Health and Safety in Automation and Robotics,* ed. K. Noro (Chicago: National Safety Council, 1987), p. 383.
25. Ibid., p. 384.

26. Goto, M. "Occupational Safety and Health Measures Taken for the Introduction of Robots in the Automobile Industry," in *Occupational Health and Safety in Automation and Robotics,* ed. K. Noro (Chicago: National Safety Council, 1987), pp. 399–417.

27. Ibid., pp. 404–408.

28. Ibid., pp. 411–13.

Bloodborne Pathogens

Acquired immunodeficiency syndrome, or AIDS, has become one of the most difficult issues that safety and health professionals are likely to face today. It is critical that they know how to deal properly and appropriately with this controversial disease. The major concerns of safety and health professionals with regard to AIDS are knowing the facts about AIDS; knowing the legal concerns associated with AIDS; knowing their role in AIDS education and related employee counseling; and knowing how to ease unfounded fears concerning the disease while simultaneously taking the appropriate steps to protect employees from infection. In addition to AIDS, the modern safety and health manager must be concerned with other bloodborne pathogens including human immunodeficiency virus (HIV) and hepatitis B (HBV).

SYMPTOMS OF AIDS

AIDS and various related conditions are caused when humans become infected with the human immunodeficiency virus, or HIV. This virus attacks the human immunity system, rendering the body incapable of repelling disease-causing microorganisms. Symptoms of the onset of AIDS are as follows:

- Enlarged lymph nodes that persist
- Persistent fevers
- Involuntary weight loss
- Fatigue
- Diarrhea that does not respond to standard medications
- Purplish spots or blotches on the skin or in the mouth
- White, cheesy coating on the tongue
- Night sweats
- Forgetfulness

How AIDS Is Transmitted

The HIV virus is transmitted in any of the following three ways: (1) sexual contact, (2) blood contact, and (3) mother-to-child during pregnancy or childbirth. Any act in which body fluids are exchanged can result in infection if either partner is infected. The following groups of people are at the highest level of risk with regard to AIDS: (1) homosexual men who do not take appropriate precautions; (2) IV drug users who share needles; (3) people with a history of multiple blood transfusions or blood-product transfusions, such as hemophiliacs; and (4) sexually promiscuous people who do not take appropriate precautions.

How AIDS Is Not Transmitted

There is a great deal of misunderstanding about how AIDS is transmitted. This can cause inordinate fear among fellow employees of HIV-positive workers. Safety and health professionals should know enough about AIDS transmission so that they can reduce employees' fears about being infected through casual contact with an HIV-positive person.

Occupational Health and Safety magazine provides the following clarifications concerning how AIDS is *not* transmitted:

> AIDS is a blood-borne, primarily sexually transmitted disease. It is not spread by casual social contact in schools, workplaces, public washrooms, or restaurants. It is not spread via handshakes, social kissing, coughs, sneezes, drinking fountains, swimming pools, toilet facilities, eating utensils, office equipment, or by being next to an infected person.
>
> No cases of AIDS have been reported from food being either handled or served by an infected person in an eating establishment.
>
> AIDS is not spread by giving blood. New needles and transfusion equipment are used for every donor.
>
> AIDS is not spread by mosquitoes or other insects.
>
> AIDS is not spread by sexual contact between uninfected individuals—whether homosexual or heterosexual—if an exclusive sexual relation has been maintained.[1]

AIDS IN THE WORKPLACE

The first step in dealing with AIDS at the company level is to develop a comprehensive AIDS policy. Safety and health professionals should be part of the team that drafts the initial policy and updates an existing policy. If a company has no AIDS policy, the safety and health professional should encourage the company to develop one. In all likelihood, most companies won't take much convincing.

According to Peter Minetos,

> Industry is doing its part to eliminate any unnecessary fear: Nearly half of companies surveyed offer their employees literature or other materials to keep them informed on the disease; more than half have an Employee Assistance Program (EAP) to deal with emotional problems concerning AIDS; two-thirds of those who have not yet addressed AIDS with employees plan to do so in the future.[2]

AIDS is having a widely felt impact in the workplace, particularly on employers. According to Minetos, employers are feeling the impact of AIDS in increased insurance premiums and health-care costs, time-on-the-job losses, decreased productivity, AIDS-related lawsuits, increased stress, and related problems that result from misconceptions about AIDS.[3]

The starting point for dealing with AIDS in the workplace is the development of a company policy that covers AIDS and other bloodborne pathogens. The policy should cover the following areas at a minimum: employee rights, testing, and education (Figure 14–1).

Figure 14–1
Components of a corporate policy for bloodborne pathogens.

**Checklist of the
Components of a Corporate Policy**

✓ Employee rights

✓ Testing

✓ Education

Employee Rights

An AIDS policy should begin by spelling out the rights of employees who have tested positive for the disease. The report *National Academy of Sciences Confronting AIDS: Update* makes the following recommendations for developing the employee rights aspects of an AIDS policy:

- Treat HIV-positive employees compassionately, allowing them to work as long as they are able to perform their jobs.

- Develop your company's AIDS policy and accompanying program before learning that an employee is HIV positive. This will allow the company to act instead of having to react.

- Make reasonable allowances to accommodate the HIV-positive employee. The U.S. Supreme Court has recognized AIDS as a handicapping condition. Consequently, reasonable allowances must include modified work schedules and special adaptations to the work environment.

- Ensure that HIV-positive employees have access to private health insurance that covers the effects of AIDS. Also, work with state and federal government insurance providers to gain their support in helping to cover the costs of health care for HIV-positive employees.

- Include provisions for evaluating the work skills of employees to determine if there has been any degradation of ability caused by the disease.[4]

Testing

According to the Centers for Disease Control, there is no single test that can reliably diagnose AIDS.[5] However, there is a test that can detect antibodies produced in the blood to fight the virus that causes AIDS. The presence of these antibodies does not necessarily mean that a person has AIDS. According to the Centers for Disease Control,

> Presence of HTLV-III antibodies [now called HIV antibodies] means that a person has been infected with that virus . . . The antibody test is used to screen donated blood and plasma and assist in preventing cases of AIDS resulting from blood transfusions or use of blood products, such as Factor VIII, needed by patients with hemophilia. For people who think they may be infected and want to know their health status, the test is available through private physicians, most state or local health departments and at other sites. Anyone who tests positive should be considered potentially capable of spreading the virus to others.[6]

Whether a company can, or even should, require AIDS tests of employees or potential employees is widely debated. The issue is contentious and controversial. However, there is a growing body of support for mandatory testing. The legal and ethical concerns surrounding this issue are covered in the next section. The testing component of a company's AIDS policy should take these concerns into account.

Education

The general public is becoming more sophisticated about AIDS. People are beginning to learn how AIDS is transmitted. However, research into the causes, diagnosis, treatment, and prevention of this disease is ongoing. The body of knowledge changes continually. Consequently, it is important to have an ongoing education program to keep employees up to date and knowledgeable. According to Minetos,

> AIDS education campaigns can be conducted in many forms. Literature, slide shows, and video presentations are all communication vehicles. Presentations by health professionals are one of the most popular and effective methods of communicating information on AIDS. The primary purpose of each is to convey basic knowledge and, subsequently, eliminate unnecessary fear among co-workers.[7]

Once a comprehensive AIDS policy has been developed and shared with all employees, a company has taken the appropriate and rational approach for dealing with this deadly and controversial disease. If an employer has not yet taken this critical step, safety and health professionals should encourage such action immediately. It is likely that most companies either employ now, or will employ in the future, HIV-positive employees. A poll conducted by *U.S. News and World Report* found that 48 percent of the companies responding indicated that AIDS was a concern.[8]

LEGAL CONCERNS

There are legal considerations relating to AIDS in the workplace with which safety and health professionals should be familiar. They grow out of several pieces of federal legislation, including the Rehabilitation Act of 1973, the Occupational Safety and Health Act of 1970, and the Employee Retirement Income System Act of 1974.

The Rehabilitation Act of 1973 was enacted to give protection to people, including workers, with handicaps. Section 504 of the act makes discrimination on the basis of a handicap unlawful. Any agency, organization, or company that receives federal funding falls within the purview of the act. Such entities may not discriminate against individuals who are handicapped but otherwise qualified. Through various court actions, this concept has been well defined. A person with a handicap is "otherwise qualified" when he or she can perform what the courts have described as the essential functions of the job.

When the handicap that a worker has is a contagious disease such as AIDS, it must be shown that there is no significant risk of the disease being transmitted in the workplace. If there is a significant risk, the infected worker is not considered otherwise qualified. Employers and the courts must make these determinations on a case-by-case basis.

Another concept associated with the Rehabilitation Act is the concept of reasonable accommodation. In determining if a worker with a handicap can perform the essential functions of a job, employers are required to make reasonable accommodations to help the worker. This concept applies to workers with any type of handicapping condition including a communicable disease such as AIDS. What constitutes reasonable accommodation, just as what constitutes otherwise qualified, must be determined on a case-by-case basis.

The concepts growing out of the Rehabilitation Act of 1973 give the supervisor added importance when dealing with AIDS-infected employees. The supervisor's knowledge of the various jobs in his or her unit will be essential in helping company officials make an otherwise qualified decision. The supervisor's knowledge that AIDS is transmitted only by exchange of body fluids coupled with his or her knowledge of the job tasks in question will be helpful in determining the likelihood that AIDS may be transmitted to other employees. Finally, the supervisor's knowledge of the job tasks in question will be essential in determining what constitutes reasonable accommodation and what the actual accommodations should be. Therefore, it is critical that safety and health professionals work closely with supervisors and educate them in dealing with AIDS in the workplace.

In arriving at what constitutes reasonable accommodation, employers are not required to make fundamental changes that alter the nature of the job or result in undue costs or administrative burdens. Clearly, good judgment and a thorough knowledge of the job are required when attempting to make reasonable accommodations for an

AIDS-infected employee. Safety and health professionals should involve supervisors in making such judgments.

The Occupational Safety and Health Act of 1970 (OSHAct) requires that employers provide a safe workplace free of hazards. The act also prohibits employers from retaliating against employees who refuse to work in an environment they believe may be unhealthy (Section 654). This poses a special problem for employers of AIDS-infected employees. Other employees may attempt to use Section 654 of the OSHAct as the basis for refusing to work with such employees. For this reason, it is important that companies educate their employees about AIDS and how it is transmitted. If employees know how AIDS is transmitted, they will be less likely to exhibit an irrational fear of working with an infected colleague.

Even when a comprehensive AIDS education program is provided, employers should not automatically assume that an employee's fear of working with the infected individual is irrational. Employers have an obligation to treat each case individually. Does the complaining employee have a physical condition that puts him or her at greater risk of contracting AIDS than other employees? If so, that employee's fears may not be irrational. However, a fear of working with an AIDS-infected co-worker is usually irrational, making it unlikely that Section 654 of the OSHAct could be used successfully as the basis of a refusal to work.

The Employee Retirement Income Security Act (ERISA) of 1974 protects the benefits of employees by prohibiting actions taken against them based on their eligibility for benefits. This means that employers covered by ERISA cannot terminate an employee with AIDS or who is suspected of having AIDS as a way of avoiding expensive medical costs. With ERISA, it is irrelevant whether the employee's condition is considered a handicap since the act applies to all employees regardless of condition.

The Testing Issue

Perhaps the most contentious legal concern growing out of the AIDS controversy is the issue of testing. Writing in the *AAOHN Journal*, Beatrice Crofts Yorker says,

> Few topics have generated the amount of controversy that currently exists in the area of testing for Acquired Immune Deficiency Syndrome (AIDS). Proponents and opponents have strong arguments, often based on emotional reactions to this deadly epidemic. In the workplace, the issues of AIDS testing are very specific and have implications for health policies in occupational settings. Few clear laws or statutes specifically regulate AIDS testing in the workplace.[9]

Issues regarding testing for AIDS and other diseases with which safety and health professionals should be familiar are summarized as follows (see Figure 14–2):

1. *State laws.* Control of communicable diseases is typically considered to be the province of the individual state. In response to the AIDS epidemic, several states have passed legislation. Some states prohibit the use of pre-employment AIDS tests to deny

Figure 14–2
Disease testing issues.

Checklist of Disease Testing Issues

Health and safety professionals should be familiar with how the following factors might affect the issue of AIDS testing at their companies:

✓ Applicable state laws

✓ Applicable federal laws and regulations

✓ Case law from civil suits

✓ Company policy

employment to infected individuals. Because of the differences among states with regard to AIDS-related legislation, safety and health professionals should familiarize themselves with the laws of the state in which their company is located.

2. *Federal laws and regulations.* The laws protecting an individual's right to privacy and due process apply to AIDS testing. These laws fall within the realm of constitutional law. They represent the primary federal contribution to the testing issue.

3. *Civil suits.* Case law serves the purpose of establishing precedents that can guide future decisions. One precedent-setting case was taken all the way to the Supreme Court (*School Board of Nassau County v. Arline*, 1987), where it was decided that an employer cannot discriminate against an employee who has a communicable disease.

4. *Company policy.* It was stated earlier that companies should have an AIDS policy that contains a testing component. This component should include at least the following: a strong rationale, procedures to be followed, employee groups to be tested, the use and dissemination of results, and the circumstances under which testing will be done. Safety and health professionals should be knowledgeable about their company's policy and act in strict accordance with it.[10]

On one side of the testing controversy are the issues of fairness, accuracy, and confidentiality or, in short, the rights of the individual. On the other side of the controversy are the issues of workplace safety and public health. Individual rights' proponents ask such questions as: What tests will be used? How do test results relate to the maintenance of workplace safety? How will test results be used and who will see them? Workplace safety proponents ask such questions as: What is the danger of transmitting the disease to other employees? Can the safety of other workers be guaranteed?

Bayer, Levine, and Wolf recommend the following guidelines for establishing testing programs that satisfy the concerns of both sides of the issue:

- The purpose of screening must be ethically acceptable.
- The means to be used in the screening program and the intended use of the information must be appropriate for accomplishing the purpose.
- High-quality laboratory services must be used.
- Individuals must be notified that the screening will take place.
- Individuals who are screened must have a right to be informed of the results.
- Sensitive and supportive counseling programs must be available before and after screening to interpret the results, whether they be positive or negative.
- The confidentiality of screened individuals must be protected.[11]

Facts About Testing for AIDS and Other Diseases

In addition to understanding the legal concerns associated with disease testing, safety and health professionals should also be familiar with the latest facts about AIDS tests and testing. A concerned employee might ask for recommendations concerning AIDS testing.

Ensuring the accuracy of an HIV antibody test (there is no such thing as an AIDS test) requires two different tests, one for initial screening and one for confirmation. The screening test currently used is the enzyme linked immunosorbent assay, or ELISA test. The confirmation test is the immuno-florescent (IFA), or the Western Blot test. The ELISA test is relatively accurate, but it is susceptible to both false positive and false negative results. A false positive test is one that shows the presence of HIV antibodies when in reality no such antibodies exist. A false negative result is one that shows no HIV antibodies in people who in reality are infected. A negative result indicates that no infection exists at the time of the test. A confirmed positive result indicates that HIV antibodies are present in the blood.[12]

The American College Health Association makes the following recommendations concerning the HIV antibody test:

1. The test is not a test for AIDS, but a test for antibodies to HIV, the virus that can cause AIDS.
2. Talk to a trained, experienced, sensitive counselor before deciding whether to be tested.
3. If you decide to be tested, do so *only* at a center that provides both pre- and post-test counseling.
4. If possible, use an *anonymous* testing center.
5. Be sure that the testing center uses two ELISA tests and a Western Blot or IFA test to confirm a positive result.
6. A positive test result is *not* a diagnosis of AIDS. It does mean you have HIV infection and that you should seek medical evaluation and early treatment.
7. A positive test result *does* mean that you can infect others and that you should avoid risky or unsafe sexual contact and IV needle sharing.
8. It can take six months (and—although rarely—sometimes even longer) after infection to develop antibodies, so the test result may not indicate whether you have been infected during that period.
9. A negative test result *does not* mean that you are immune to HIV or AIDS, or that you cannot be infected in the future.[13]

Safety and health professionals need to share this type of information with employees who ask about AIDS tests and testing. Employees who need more detailed information should be referred to a health-care professional.

AIDS EDUCATION

The public is becoming more knowledgeable about AIDS and how the disease is spread. However, this is a slow process, and AIDS is a complex and controversial disease. Unfortunately, many people still respond to the disease out of ignorance and inaccurate information. For this reason, it is imperative that a company's safety and health program include an AIDS education program.

A well-planned AIDS education program can serve several purposes: (1) It can give management the facts needed to develop policy and make informed decisions with regard to AIDS; (2) it can result in changes in behavior that will make employees less likely to contract or spread the disease; (3) it can prepare management and employees to respond appropriately when a worker falls victim to the disease; and (4) it can decrease the likelihood of legal problems resulting from an inappropriate response to an AIDS-related issue. Consequently, safety and health professionals should be prepared to participate in developing AIDS education programs.

Planning an AIDS Education Program

The first step in planning an AIDS education program is to decide its purpose. A statement of purpose for an AIDS education program should be a broad conceptual declaration that captures the company's reason for providing the education program. Following is an example:

> The purpose of this AIDS education program is to deal with the disease in a positive proactive manner that is in the best interests of the company and its employees.

The next step in the planning process involves developing goals that translate the statement of purpose into more specific terms. The goals should tell specifically what the AIDS education program will do. Sample goals are as follows:

- The program will change employee behaviors that might otherwise promote the spread of AIDS.
- The program will help the company's management team to develop a rational, appropriate AIDS policy.

Figure 14–3
Course outline for AIDS education course.

Statement of Purpose

The purpose of this course is to give employees the knowledge they need to deal with AIDS
in a positive, proactive manner.

Major Topics

- What is AIDS?
- What causes AIDS?
- How is AIDS transmitted?
- Who is most likely to get AIDS?
- What are the symptoms of AIDS?
- How is AIDS diagnosed?
- Who should be tested for AIDS?
- Where can I get an AIDS test?
- How can I reduce my chances of contract-
  ing AIDS?
- How is AIDS treated?
- Can AIDS be prevented?
- What are common myths about AIDS?

- The program will help managers to make responsible decisions concerning AIDS issues.
- The program will help employees to protect themselves from the transmission of AIDS.
- The program will alleviate the fears of employees concerning working with an AIDS-
  infected co-worker.
- The program will help managers to respond appropriately and humanely to the
  needs of AIDS-infected workers.

Once goals have been set, a program is developed to meet the goals. The various
components of the program must be determined. These components may include confidential one-on-one counseling, referral, posters, a newsletter, classroom instruction, self-
paced multimedia instruction, group discussion sessions, printed materials, or a number
of other approaches. Figure 14–3 is a suggested outline for a course on AIDS.

PROTECTING EMPLOYEES FROM AIDS

Safety and health professionals should be familiar with the precautions that will protect
employees from HIV infection on and off the job. OSHA's guidelines for preventing exposure to HIV infection identify three categories of work-related tasks: Categories I, II, and
III. Jobs that fall into Category I involve routine exposure to blood, body fluids, or tissues
that might be HIV-infected. Category II jobs do not involve routine exposure to blood,
body fluids, or tissues, but some aspects of the job may involve occasionally performing
Category I tasks. Category III jobs do not normally involve exposure to blood, body fluids, or tissues.

Most industrial occupations fall into Category III, meaning there is very little risk of
contracting AIDS on the job. However, regardless of the category of their job, employees
should know how to protect themselves, and safety and health professionals should be
prepared to tell them how.

The U.S. Public Health Service recommends the following precautions for reducing
the chances of contracting AIDS:

- Abstain from sex or have a mutually monogamous marriage/relationship with an infection-free partner.
- Refrain from having sex with multiple partners or with a person who has multiple partners. The more partners one has, the greater the risk of infection.
- Avoid sex with a person who has AIDS or who you think might be infected. However, if you choose not to take this recommendation, the next logical course is to take precautions against contact with the infected person's body fluids (blood, semen, urine, feces, saliva, and female genital secretions).
- Do not use intravenous drugs or, if you do, do not share needles.[14]

Safety and health professionals should make sure that all employees are aware of these common precautions. They should be included in the company-sponsored AIDS education program, available through the company's EAP, and posted conspicuously for employee reading.

CPR and AIDS

It is not uncommon for an employee to be injured in a way that will require resuscitation. Consequently, many companies provide employees with CPR training. But what about AIDS? Is CPR training safe? Writing for *Safety and Health*, Martin Eastman said,

> In the early 1960s when CPR training procedures were being developed, some thought was given to the cleaning and disinfection of manikins. This was before acquired immune deficiency syndrome (AIDS) and hepatitis-B became headline diseases. In the fearful climate that has followed the spread of these diseases practices that once appeared merely unsanitary now seem truly life-threatening.[15]

The HIV virus has been found in human saliva. Because CPR involves using your fingers to clear the airway and placing your mouth over the victim's, there is concern about contracting AIDS while trying to resuscitate someone. Although there is no hard evidence that HIV can be transmitted through saliva, there is some legitimacy to the concern.

Because of the concern, disposable face masks and various other types of personal protective devices are now being manufactured.

Safety and health professionals should ensure that such devices are used in both training and live situations involving CPR. These devices should be readily available in many easily accessible locations throughout the company.

HEPATITIS B (HBV) IN THE WORKPLACE

Although the spread of HIV receives more attention, a greater risk is from the spread of Hepatitis B. This bloodborne virus averages approximately 300,000 new cases per year compared with AIDS which averages approximately 35,000 new cases. The Hepatitis B virus is extremely strong as compared with HIV. For example, it can live on surfaces for up to a week if it is exposed to air. Hepatitis B is also much more concentrated than HIV.

Hepatitis B is caused by a double-shelled virus (HBV). It can be transmitted in the workplace in the following ways:

- Contact with blood
- Contact with bodily fluids including tears, saliva, and semen

The Hepatitis B virus can live in bodily fluids for years. Carriers of the virus are at risk themselves, and they place others at risk. Persons infected with HBV may contract chronic hepatitis, cirrhosis of the liver, and/or primary heptocellular carcinoma. An HBV-infected individual is more than 300 times more likely to develop primary liver cancer than is a noninfected individual from the same environment. Unfortunately, it is possible to be infected and not know it because the symptoms can vary so widely from person to person.

The symptoms of Hepatitis B are varied but include the following:

- Jaundice
- Joint pain
- Rash
- Internal bleeding

The next section explains OSHA's standard on occupational exposure to bloodborne pathogens. This standard applies to all bloodborne pathogens including HBV and HIV.

OSHA'S STANDARD ON OCCUPATIONAL EXPOSURE TO BLOODBORNE PATHOGENS

OSHA's standard on occupational exposure to bloodborne pathogens is contained in 29 C.F.R. Part 1910.1030. The purpose of the standard is to limit the exposure of personnel to blood and to serve as a precaution against bloodborne pathogens that may cause diseases.

Scope of Application

This standard applies to all employees whose job duties may bring them in contact with blood or other potentially infectious material. There is no attempt on OSHA's part to list all occupations to which 1910.1030 applies. The deciding factor is the *reasonably anticipated* theory. If it can be reasonably anticipated that employees may come in contact with blood in the normal course of performing their job duties, the standard applies. It does not apply in instances of *good Samaritan* type acts in which one employee attempts to assist another employee who is bleeding.

The standard speaks to blood and other infectious materials. These other materials include the following:

- Semen
- Vaginal secretions
- Cerebrospinal fluid
- Synovial fluid
- Pleural fluid
- Peritoneal fluid
- Amniotic fluid
- Saliva
- Miscellaneous body fluids mixed with blood

In addition to these fluids, other potentially infectious materials include the following:

- Unfixed human tissue, or organs other than intact skin
- Cell or tissue cultures
- Organ cultures
- Any medium contaminated by Human Immunodeficiency Virus (HIV) or Hepatitis B (HBV)

Exposure Control Plan

OSHA's 1910.1030 requires employers to have an *Exposure Control Plan* to protect employees from exposure to bloodborne pathogens. It is recommended that such plans have at least the following major components:

Part 1   Administration
Part 2   Methodology

Part 3	Vaccinations
Part 4	Post-Exposure Investigation and Follow-Up
Part 5	Labels and Signs
Part 6	Information and Training

Administration

This component of the plan should clearly define the responsibilities of employees, supervisors, and managers regarding exposure control. It should also designate an exposure control officer (usually the organization's safety and health manager or a person who reports to this manager). The organization's Exposure Control Plan must be readily available to all employees, and the administration component of it must contain a list of locations where copies of the plan can be examined by employees. It should also describe the responsibilities of applicable constituent groups/individuals. These responsibilities are summarized in the following sections.

Employees   All employees are responsible for the following: knowing which of their individual and specific job tasks may expose them to bloodborne pathogens; participating in training provided concerning bloodborne pathogens; carrying out all job duties in accordance with the organization's various control procedures; and practicing good personal hygiene.

Supervisors and Managers   Supervisors and managers are responsible for coordinating with the exposure control officer to implement and monitor exposure control procedures in their areas of responsibility.

Exposure Control Officer   This individual is assigned overall responsibility for carrying out the organization's exposure control plan. In addition to the administrative duties associated with it, this position is also responsible for employee training. In small organizations, exposure control officers may double as the training coordinator. In larger organizations, a separate training coordinator may be assigned exposure control as one more training responsibility.

In either case, the exposure control officer is responsible for the following duties: overall development and implementation of the exposure control plan; working with management to develop other exposure-related policies; monitoring and updating the plan; keeping up-to-date with the latest legal requirements relating to exposure; acting as liaison with OSHA inspectors; maintaining up-to-date training files and documentation showing training; developing the needed training program; working with other managers to establish appropriate control procedures; establishing a Hepatitis B vaccination program as appropriate; establishing a post-exposure evaluation and follow-up system; displaying labels and signs as appropriate; and maintaining up-to-date, confidential medical records of exposed employees.

Methodology

This section of the plan describes the procedures established to protect employees from exposure. These procedures fall into one of the following five categories:

- General precautions
- Engineering controls
- Work practice controls
- Personal protection equipment
- Housekeeping controls

General precautions include such procedures as assuming that all body fluids are contaminated, and acting accordingly. Engineering controls are design and technological precautions that protect employees from exposure. Examples of engineering controls include self-sheathing needles, readily accessible hand-washing stations equipped with antiseptic hand cleaners, leakproof specimen containers, and puncture-proof containers for sharp tools that are reusable. Work practice controls are precautions that individual employees take, such as washing their hands immediately after removing potentially contaminated gloves or refraining from eating or drinking in areas where bloodborne pathogens may be present. Personal protection equipment includes any device designed to protect an employee from exposure. Widely used devices include the following:

- Gloves (disposable and reusable)
- Goggles and face shields
- Respirators
- Aprons, coats, and jackets

Examples of housekeeping controls include the use, disposal, and changing of protective coverings; decontamination of equipment; and regular cleaning of potentially contaminated areas.

Vaccinations

The OSHA standard requires that employers make Hepatitis B vaccinations available to all employees for whom the *reasonable anticipation* rule applies. The vaccination procedure must meet the following criteria: available at no cost to employees; administered at a reasonable time and place within ten days of assignment to a job with exposure potential; and administered under the supervision of an appropriately licensed health-care professional, according to the latest recommendations of the U.S. Public Health Service.

Employees may decline the vaccination, but those who do must sign a *declination form* stating that they understand the risk to which they are subjecting themselves. Employees who decline are allowed to change their minds. Employees who do must be allowed to receive the vaccination.

Part of the exposure control officer's job is to keep accurate, up-to-date records showing vaccinated employees, vaccination dates, employees who declined, and signed declination forms.

Post-Exposure Investigation and Follow-Up

When an employee is exposed to bloodborne pathogens, it is important to evaluate the circumstances and follow up appropriately. How did it happen? Why did it happen? What should be done to prevent future occurrences? The post-exposure investigation should determine at least the following:

- When did the incident occur?
- Where did the incident occur?
- What type of contaminated substances or materials were involved?
- What is the source of the contaminated materials?
- What type of work tasks were being performed when the incident happened?
- What was the cause of the incident?
- Were the prescribed precautions being observed when the incident occurred?
- What immediate action was taken in response to the incident?

Using the information collected during the investigation, an incident report is written. This report is just like any other accident report. Keeping the exposed employee fully informed is important. The employee should be informed of the likely avenue of exposure and the source of the contaminated material, even if the source is another employee. If the source is another employee, that individual's blood should be tested for HBV or HIV, and the results should be shared with the exposed employee. Once the exposed employee is fully informed, he or she should be referred to an appropriately certified medical professional to discuss the issue. The medical professional should provide a written report to the employer containing all pertinent information and recommendations. The exposed employee should also receive a copy.

Labels and Signs

This section of the plan describes the procedures established for labeling potential biohazards. Organizations may also use warning signs as appropriate and color-coded containers. It is important to label or designate with signs the following:

- Biohazard areas
- Contaminated equipment
- Containers of potentially contaminated waste
- Containers of potentially contaminated material (i.e., a refrigerator containing blood)
- Containers used to transport potentially contaminated material

Information and Training

This section of the plan describes the procedures for keeping employees knowledgeable, fully informed, and up-to-date regarding the hazards of bloodborne pathogens. The key to satisfying this requirement is training. Training provided should cover at least the following:

- OSHA Standard 1910.1030
- The exposure control plan
- Fundamentals of bloodborne pathogens (e.g., epidemiology, symptoms, and transmission)
- Hazard identification
- Hazard prevention methods
- Proper selection and use of personal protection equipment
- Recognition of warning signs and labels
- Emergency response techniques
- Incident investigation and reporting
- Follow-up techniques
- Medical consultation

It is important to document training and keep accurate up-to-date training records on all employees. These records should be available to employees and their designated representatives (e.g., family members, attorneys, and physicians), and to OSHA personnel. They must be kept for at least three years and must include the following:

- Dates of all training
- Contents of all training
- Trainers' names and qualifications
- Names and job titles of all participants

Record Keeping

OSHA Standard 1910.1030 requires that medical records be kept by employers on all employees who are exposed to bloodborne pathogens. These records must be confidential and should contain at least the following information:

- Employee's name and social security number
- Hepatitis B vaccination status
- Results of medical examinations and tests
- Results of incident follow-up procedures

- *Acid rain.*   Gradually reduce sulfur oxide emissions while concurrently avoiding expensive new regulation-induced clampdowns until sufficient evidence is available to justify such actions.
- *Ground-level ozone.*   Develop new regulations that focus on gas stations and other small sources of ground-level ozone emissions while concurrently avoiding any new drastic restrictions on automobile emissions.
- *Global warming.*   Focus more on the use of nuclear power while continuing efforts to use fossil fuels more efficiently.
- *Water pollution.*   Build more sewage treatment plants throughout the nation as quickly as possible.
- *Toxic wastes.*   Develop and implement incentive programs to encourage a reduction in the volume of toxic waste.
- *Garbage.*   Increase the use of recycling while simultaneously reducing the overall waste stream. Burn or bury what cannot be recycled, but under strict controls.[3]

LEGISLATION AND REGULATION

A fairly clear-cut division of authority exists for legislation and regulations concerning the environment. The Occupational Safety and Health Administration (OSHA) is responsible for regulating the work environment within an individual company or plant facility. Environmental issues that go beyond the boundaries of the individual plant facility are the responsibility of the EPA. Of course, some environmental issues and concerns do not fall clearly within the scope of either OSHA or the EPA. Therefore, these two agencies have begun to cooperate closely in dealing with environmental matters.

Clean Air Act

One of the most important pieces of federal environmental legislation has been the Clean Air Act. Originally passed in 1970, the act was amended in 1990. Writing in the magazine *Occupational Hazards*, Greg LaBar had this to say about the amended Clean Air Act:

> Relying on technology-driven, market-based strategies, the law is designed to reduce air pollution—in the form of hazardous air pollutants, acid rain, and smog—by 56 billion lb. per year. This includes a 75 percent reduction in air toxins, a 50 percent cut in acid rain, and 40 percent decrease in smog over the next 20 years or so. Like the original Clean Air Act (CAA) of 1970, the new law does not on its own guarantee clean air, or even cleaner air. It's a framework from which the Environmental Protection Agency (EPA) and state agencies will develop the implementing regulations.[4]

The Clean Air Act contains provisions that require companies to take whatever actions are necessary to prevent or minimize the potential consequences of the accidental release of pollutants into the air. It also establishes an independent chemical safety and hazard investigation board that will investigate accidental releases of pollutants that result in death, serious injury, or substantial property damage.

The Clean Air Act as amended in 1990 contains approximately 350 pages detailing the requirements of seven titles:

Title I:	Urban Air Quality
Title II:	Mobile Sources
Title III:	Hazardous Air Pollutants
Title IV:	Acid Rain Control
Title V:	Permits

Title VI: Stratospheric Ozone Provisions

Title VII: Enforcement

Figure 15–1 gives a brief summary of each title in the act.

HAZARDOUS WASTE REDUCTION

One of the most effective ways to ensure a safe and healthy environment is to reduce the amount of hazardous waste being introduced into it. In 1976, the EPA developed a policy statement designed to encourage industrial firms to reduce their hazardous waste output. The policy contains three broad provisions. First, companies must certify on the transportation manifest that they have a program in operation to reduce the volume and toxicity of hazardous waste each time hazardous waste is transported off-site. Second, in order to qualify for permits to treat, store, and/or dispose of hazardous waste, companies must implement and operate a hazardous waste reduction program. Finally, companies must submit biennial plans to the EPA describing actions taken to reduce the volume and toxicity of their hazardous waste.

In 1976, the Resource Conservation and Recovery Act (RCRA) became the first piece of federal legislation to encourage the reduction of hazardous waste. It was followed in 1984 by passage of the Hazardous and Solid Waste Amendments (HSWA). However, the most significant federal regulation pertaining to hazardous waste reduction is the OSHA Hazardous Waste Standard, which took effect in 1990.

OSHA Hazardous Waste Standard

On March 6, 1990, OSHA standard (1910.120) went into effect setting standards for dealing with hazardous materials.

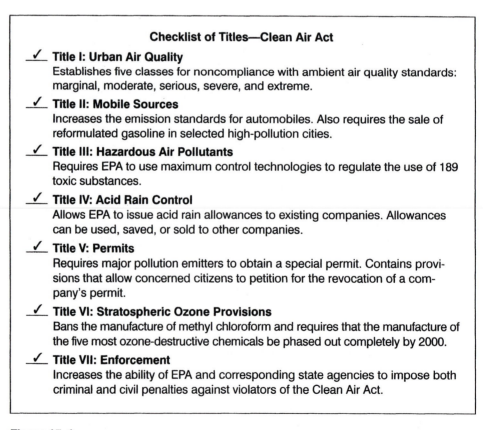

Figure 15–1
Clean Air Act.

> **Checklist of the General Requirements—OSHA Hazardous Waste Standard**
>
> ✓ Develop a safety and health program to control hazards and provide for emergency response.
>
> ✓ Conduct a preliminary site evaluation to identify potential hazards and select appropriate employee protection strategies.
>
> ✓ Implement a site control program to prevent contamination of employees.
>
> ✓ Train employees before allowing them to undertake hazardous waste operations or emergency response activities that could expose them to hazards.
>
> ✓ Provide medical surveillance at least annually and/or at the end of employment for employees exposed to greater than acceptable levels of specific substances.
>
> ✓ Implement measures to reduce exposure to hazardous substances to below established acceptable levels.
>
> ✓ Monitor air quality to identify and record the levels of hazardous substances in the air.
>
> ✓ Implement a program to inform employees of the names of people responsible for health and safety and of the requirements of the OSHA standard.
>
> ✓ Implement a decontamination process that is used each time an employee or piece of equipment leaves a hazardous area.
>
> ✓ Develop an emergency response plan to handle on-site emergencies.
>
> ✓ Develop an emergency response plan to coordinate off-site services.

Figure 15–2
General requirements of the OSHA Hazardous Waste Standard.

Figure 15–2 summarizes the requirements of the OSHA Hazardous Waste Standard. These standards require that companies identify, evaluate, and control hazardous substances; train employees in the proper accident prevention procedures and emergency response procedures; reduce exposure to hazardous substances; provide medical monitoring; keep employees informed concerning hazardous substances on the job; implement decontamination procedures; and develop both on- and off-site emergency response plans.

The goal of these and other hazardous materials regulations is to (1) encourage companies to minimize the amount and toxicity of hazardous substances that they use; (2) ensure that all remaining hazardous materials are used safely; and (3) ensure that companies are prepared to respond promptly and appropriately when accidents occur.

Safety and health professionals need to be familiar with the concept of hazardous waste reduction, how to organize a reduction program, and how to conduct a reduction audit. Hazardous waste reduction can be defined as follows:

The two key elements of this definition are source reduction and recycling. Figure 15–3 breaks down these two elements into more specific subelements.

> *Hazardous waste reduction is reducing the amount of hazardous waste generated and, in turn, the amount introduced into the waste stream through the processes of source reduction and recycling.*

Organizing a Waste Reduction Program

There are four steps that must be accomplished in establishing a waste reduction program. The first step is to convince top-level managers that the program is not just envi-

Figure 15–3
Key elements in hazardous waste reduction.

Checklist of Key Elements in Hazardous Waste Reduction

Source Reduction

✓ Management improvements in operating efficiency

✓ Better use of modern technology and processes

✓ Better selection of the materials used in production processes

✓ Product revisions

Recycling

✓ Reclamation of useable materials from hazardous waste materials

✓ Reconstituting of waste for reuse as an original product

---

ronmentally and ethically necessary, but also cost effective. The cost/benefit data associated with waste reduction should be enough to convince the most cost-conscious manager of the feasibility of such a program. The potential for reduced costs in the areas of regulatory compliance, legal liability, and workers' compensation should be included in all briefings given to managers. The other steps involved in organizing a waste reduction program are discussed in the following paragraphs (see Figure 15–4). Figure 15–5 is a checklist that can be used when developing a comprehensive waste reduction plan.

Waste Reduction Audit

The most important steps in a waste reduction audit are as follows: (1) target processes; (2) analyze processes; (3) identify reduction alternatives; (4) consider the cost/benefit ratio for each alternative; and (5) select the best options (Figure 15–6).

ISO 14000 INTRODUCED

Globalization of the marketplace has created a competitive environment that requires peak performance and continual improvement. The unrelenting demands of the modern marketplace have given rise to new philosophies for doing business.

As a result, the International Organization for Standardization (ISO)—the same organization that developed the ISO 9000 quality standards—developed the ISO 14000 family of standards to promote effective environmental management systems.

Just as the decision to adopt the ISO 9000 standards is voluntary, adoption of ISO 14000 is based on voluntary organizational commitment to environmental protection rather than government coercion. For safety and health managers accustomed to complying with government mandates, ISO 14000 certification is a novel approach.

Figure 15–4
Steps in establishing a hazardous waste reduction program.

Checklist for Hazardous Waste Reduction

✓ Gain a full commitment from the executive level of management.

✓ Form the waste reduction team.

✓ Develop a comprehensive waste reduction plan.

✓ Implement, monitor, and adjust as necessary.

Figure 15–5
Outline of waste reduction plan.

> **Checklist for Developing
> a Waste Reduction Plan**
>
> ✓ Statement of purpose
> ✓ Goals with timetables
> ✓ Strategies for accomplishing each goal
> ✓ Potential inhibitors associated with each goal
> ✓ Measures of success for each goal
> ✓ Tracking system
> ✓ Audit process

Shifting attitudes and greater public awareness are making it essential that business firms be good neighbors in their communities. The marketplace demands that businesses produce high-quality products at competitive prices, without harming the environment. ISO 14000 provides the framework for making effective environmental management part of the organization's overall management system.

WHAT IS ISO?

ISO is the acronym for International Organization for Standardization, a worldwide organization of national standards bodies (Figure 15-7). The complete membership roster for ISO contains the standards bodies of 118 countries. The overall goal of ISO is as follows:

> . . . to promote the development of standardization and related activities in the world with a view to facilitating the international exchange of goods and services and to developing cooperation in the sphere of intellectual, scientific, technological, and economic activity.[5]

ENVIRONMENTAL MANAGEMENT SYSTEM (EMS)

A management system, regardless of its application, is the component of an organization responsible for leading, planning, organizing, and controlling (Figure 15-8). An environmental management system, or EMS, is the component of an organization with primary responsibility for these functions as they relate specifically to the impact of an organization's processes, products, and/or services on the environment.

An organization's EMS may be a subset of its safety and health management system or a separate component of the organization's overall management system. Regardless of where and how it fits into an organization, the EMS should do the following:[6]

Figure 15–6
Major steps in a waste reduction audit.

> **Checklist of Major Steps Waste Reduction Audit**
>
> ✓ Target processes for analysis and create the target list.
> ✓ Analyze processes to identify sources of waste generation.
> ✓ Identify reduction options for each waste generation source.
> ✓ Consider the cost/benefit of each reduction goal.
> ✓ Select the best options.

Australia Standards Australia 1 The Crescent Homebush—N.S.W. 2140 P. O. Box 1055 Strathfield—N.S.W. 21335 Tel: +61 2 746-4700	**Canada** Standards Council of Canada 45 O'Connor Street Suite 1200 Ottawa, Ontario KIP 6N7 Tel: 1 613 238-3222
France Association francaise de normalisation Tour Europe F-92049 Paris La DeFeuse Cedex Tel: +33 1 42 91 55 55	**Germany** Deutsches Institut fur Normung Burggrafen strasse 6 D-10772 Berlin Tel: +49 30 26 01 23 44
United Kingdom British Standards Institution 389 Chiswick High Road GB-London W4 4AL Tel: +44 181 996 90 00	**United States** American National Standards Institute (ANSI) 11 West 42nd Street 13th Floor New York, N.Y. 10036 Tel: +1 212 642-4900

Figure 15–7
Addresses of selected ISO members.

- Establish a comprehensive environmental-protection policy (planning).
- Identify all government regulations and requirements that apply to the organization's processes, products, and/or services (controlling).
- Establish organization-wide commitment to environmental protection (leading).
- Establish responsibility and accountability relating to environmental protection (organizing).
- Incorporate environmental concerns in all levels of organizational planning including strategic, operational, and procedural (planning).
- Establish management processes for achieving performance benchmarks (controlling).
- Provide sufficient resources to ensure that performance benchmarks can be achieved on a continual basis (leading).
- Establish and maintain an effective emergency preparedness program (leading, planning, organizing, and controlling).
- Assess the organization's environmental performance against all applicable benchmarks and adjust as necessary (controlling).
- Establish a review process for auditing the EMS and identifying opportunities for improvement.
- Establish and maintain communications linkages with all stakeholders, internal and external.
- Promote the establishment of an EMS in contractors and suppliers.

RATIONALE FOR THE EMS MOVEMENT

Different organizations have become interested in better environmental management for different reasons (Figure 15-9). Some organizations are responding to pressure brought

Figure 15–8
The principal functions of management.

> **Checklist of Core Management Functions**
>
> *Leading*
>
> ✓ Establishing a vision
>
> ✓ Maintaining effective communication
>
> ✓ Setting an example of commitment
>
> ✓ Inspiring and motivating
>
> ✓ Providing adequate support/resources
>
> *Planning*
>
> ✓ Planning strategy
>
> ✓ Planning operation
>
> ✓ Developing policy
>
> ✓ Developing procedure
>
> *Organizing*
>
> ✓ Establishing structure
>
> ✓ Defining staff and line functions
>
> ✓ Delegating authority
>
> ✓ Establishing span of control
>
> *Controlling*
>
> ✓ Establishing benchmarks
>
> ✓ Monitoring performance
>
> ✓ Adjusting as necessary

by environmental advocacy groups and watchdog organizations. Such groups have become increasingly proficient in winning public support for their individual causes. This support, if effectively focused, can be translated into market pressure. When this happens, better environmental management becomes a market imperative.

One of the key drivers behind the EMS movement is the concept of competitive advantage. Some forward-looking organizations, particularly those that must compete in the global marketplace, have begun to view having an effective EMS as a competitive advantage. The advantage is the result of the following factors:

■ Minimization of funds siphoned off into nonproductive activities such as litigation and crisis management.

■ Ease of compliance with government regulations in the foreign countries that make up the global marketplace.

■ Better public image in countries (especially European countries) where interest in environmental protection is high.

ISO 14000 SERIES OF STANDARDS

The term ISO 14000 Series refers to a family of environmental management standards that cover the five disciplines shown in Figure 15-10. All of the standards are in a constant state of evolution. The ISO 14000 Series contains two types of standards: specification standards and guidance standards.

Figure 15–9
Forces driving the interest in EMS.

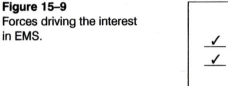

> **Checklist of Forces Driving Interest in EMS**
>
> ✓ Public pressure from environmental advocacy groups
> ✓ Influence of environmental advocates within elected bodies and governmental organizations
> ✓ Potential to gain a competitive advantage in the global marketplace
> ✓ Improved public relations
> ✓ Customer demand
> ✓ Fear of liability and risk
> ✓ Pre-empt new and additional government oversight
> ✓ Reduced duplication
> ✓ De facto requirement to adopt a standard
> ✓ Get ahead of government adoption

Specification vs. Guidance Standards

A specification standard contains only the specific criteria that can be audited internally or externally by a third party. A guidance standard explains how to develop and implement environmental management systems and principles. Guidance standards are descriptive standards that also explain how to coordinate among various quality management systems.

Guides and Technical Reports

ISO also develops guides and technical reports. A guide (not a guidance standard) is a tool to assist organizations in the improvement of environmental management. Guides are used voluntarily and do not contain criteria for certification. There is just one guide pertaining to the ISO 14000 Series: *ISO Guide 64—Guide for the Inclusion of Environmental Aspects in Product Standards.*[7]

Technical reports are written only when one of the following circumstances exist:

■ When a technical committee of ISO cannot reach consensus on an issue. In such cases, the technical report explains why consensus was not possible.

■ When an issue is still under development or when there is reason to believe that an undecided issue that cannot be immediately resolved will be resolved in the future.

■ When a technical committee that is working on a standard collects information that is different from the kind normally published as a standard. This information may be published as a technical report.

Figure 15–10
Environmental disciplines in the ISO 14000 series of standards.

> **Checklist of Environmental Disciplines**
>
> ✓ Environmental management system
> ✓ Environmental auditor criteria (These criteria may be used by internal auditors and external third-party auditors.)
> ✓ Environmental performance evaluation criteria
> ✓ Environmental labeling criteria
> ✓ Life-cycle assessment methods

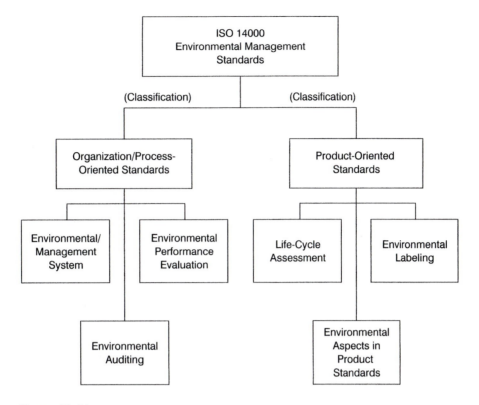

Figure 15–11
Environmental management standards.

Classification of ISO 14000 Standards

Figure 15-11 illustrates the two broad classifications of standards in the ISO 14000 family and their respective subclassifications. The two broad classifications are as follows:

- Process- or organization-oriented standards
- Product-oriented standards

    Process-oriented standards cover the following three broad areas: environmental management system, environmental performance evaluation, and environmental auditing. Product-oriented standards cover three different broad areas of concern as follows: life-cycle assessment, environmental labeling, environmental aspects in product standards.

STANDARDS IN THE ISO 14000 SERIES

The six subclassifications shown in Figure 15-11 represent six distinct, but interrelated, areas of concern. There are actually 20 separate documents in the ISO 14000 series, all related to one of these six subclassifications. Figure 15-12 is a checklist of the process-oriented documents in the ISO 14000 series. Figure 15-13 is a checklist of the product-oriented documents. Of these 20 documents, only ISO 14000 is a standard against which a company is audited. The rest are guidance documents. This is an important point to remember.

ISO 14001 STANDARD[8]

The ISO 14001 Standard represents an approach to protecting the environment and, in turn, a company whose processes and products may affect the environment. The

Figure 15–12
ISO 14000 process-oriented documents.

Checklist of ISO 14000 Process-Oriented Documents

✓ ISO 14001
Environmental Management Systems—Specification with Guidance for Use

✓ ISO 14004
Environmental Management Systems—General Guidelines on Principles, Systems, and Supporting Techniques

✓ ISO 14010
Guidelines for Environmental Auditing—General Principles on Environmental Auditing

✓ ISO 14011/1
Guidelines for Environmental Auditing—Audit Procedures—Auditing of Environmental Management Systems

✓ ISO 14012
Guidelines for Environmental Auditing—Qualification Criteria for Environmental Auditors

✓ ISO 14014
Initial Reviews

✓ ISO 14015
Environmental Site Assessments

✓ ISO 14031
Evaluation of Environmental Performance

✓ ISO 14020
Goals and Principles of All Environmental Labeling

✓ ISO 14021
Environmental Labels and Declarations—Self-Declaration Environmental Claims—Terms and Definitions

---

standard is unique because it relies on voluntary motivation instead of mandatory compliance. To understand the ISO 14001 standard, one must first understand the following concepts: environmental management system (EMS), EMS audit, environmental aspect, continual improvement, and plan-do-check-adjust model.

1. *Environmental Management System (EMS).* An organization's EMS provides the structure for implementing its environmental policy. The management system consists of the organization's structure, personnel, processes, and procedures relating to environmental management.

2. *EMS audit.* The EMS audit is the process used to verify that an EMS actually does what an organization says it will do. As part of the audit, results are reported to the organization's top management team.

3. *Environmental aspect.* Any aspect of an organization's processes, products, or services that can potentially affect the environment.

4. *Continual improvement.* In the global marketplace, good enough is never good enough. Performance that is competitive today may not be tomorrow. Consequently, an EMS must be improved continually, forever.

5. *Plan-do-check-adjust model.* The plan-do-check-adjust (PDCA) model comes from the work of J. Edwards Deming in the area of Total Quality Management. The model works well in any management system that must be continually improved. Figure 15-14 lists some of the types of activities—relating to each of the model's four components—that are associated with continually improving an EMS.

Figure 15–13
ISO 14000 product-oriented documents.

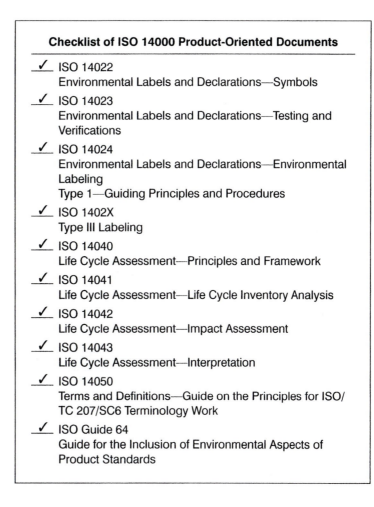

Checklist of ISO 14000 Product-Oriented Documents

✓ ISO 14022
Environmental Labels and Declarations—Symbols

✓ ISO 14023
Environmental Labels and Declarations—Testing and Verifications

✓ ISO 14024
Environmental Labels and Declarations—Environmental Labeling
Type 1—Guiding Principles and Procedures

✓ ISO 1402X
Type III Labeling

✓ ISO 14040
Life Cycle Assessment—Principles and Framework

✓ ISO 14041
Life Cycle Assessment—Life Cycle Inventory Analysis

✓ ISO 14042
Life Cycle Assessment—Impact Assessment

✓ ISO 14043
Life Cycle Assessment—Interpretation

✓ ISO 14050
Terms and Definitions—Guide on the Principles for ISO/ TC 207/SC6 Terminology Work

✓ ISO Guide 64
Guide for the Inclusion of Environmental Aspects of Product Standards

6. *Structure and application of the ISO 14001 standards.* The ISO 14001 Standard is subdivided into five broad components as shown in Figure 15-15. It is important to understand what the ISO 14001 standard does and does not do. The standard *does* allow a company voluntarily and proactively to establish a framework for involving all employees in improving environmental management. The standard does *not* establish product or performance standards, set pollutant levels, specify test methods, mandate zero emissions, nor expand on governmental regulations.

REQUIREMENTS OF THE ISO 14001 STANDARD[9]

The ISO 14001 Standard contains criteria subdivided into the following broad categories: General environmental policy; planning; implementation and operations; checking and corrective action; and management review. These broad categories are in most cases, subdivided further.

Following is a list of criteria that translates ISO 14001 into a format that can be used for conducting internal audits. The list ties into each category and subcategory of the standard.

4.0   General*

1.   Has the organization established an EMS that meets the requirements of ISO 14001?
2.   Does the organization properly maintain the EMS?

---

* Complete the remainder of the list before answering the questions in this section.

Figure 15–14
Plan-do-check-adjust (PDCA)
model applied to an EMS.

Checklist of Typical P-D-C-A Activities

Plan

✓ Identify environmental aspects of processes, products, and services

✓ Evaluate all environmental aspects identified

✓ Identify legal requirements associated with the EMS

✓ Develop an environmental policy for the organization

✓ Develop performance criteria for internal audits of the EMS

✓ Establish EMS objectives, benchmarks, and performance targets

Do

✓ Provide the resources needed to support the EMS

✓ Assign responsibility for all aspects of the EMS

✓ Clarify accountability as it relates to the performance of EMS

✓ Provide training to ensure that all parties associated with the EMS have the knowledge and skills needed

✓ Communicate continually

✓ Develop and disseminate reports

✓ Document all aspects of the EMS and its performance

✓ Establish operational control

Check

✓ Monitor the performance of the EMS by making appropriate measurements

✓ Apply preventive action wherever possible

✓ Monitor EMS records continually

✓ Conduct audits of the EMS

Act

✓ Take corrective action when necessary

✓ Implement adjustments as needed continually

4.0.1   Environmental Policy

3. Has the organization developed an appropriate environmental policy?
4. Has the organization ensured that its environmental policy meets the following criteria:
    a. Appropriate in terms of the nature, scale, and potential environmental impact of the organization's processes, products, and/or services?
    b. Contains a statement of commitment to continual improving and to preventing pollution?
    c. Contains a statement of commitment to comply with all applicable and relevant regulations (federal, state, and local)?
    d. Establishes a framework for setting environmental objectives and for monitoring progress in achieving the objectives?
    e. Is effectively deployed:
        • Documented?
        • Implementing?

Figure 15–15
Structure of the ISO 14001
Standard.

4.0	**General**
4.0.1	Policy
4.2	**Planning**
4.2.1	Environmental aspects
4.2.2	Legal and other requirements
4.2.3	Objectives and targets
4.2.4	Environmental Management programs
4.3	**Implementation and Operations**
4.3.1	Structure and Responsibility
4.3.2	Training, Awareness, and Competence
4.3.3	Communication
4.3.4	Environmental management system documentation
4.3.5	Document control
4.3.6	Operational control
4.3.7	Emergency preparedness and response
4.4	**Checking and Corrective Action**
4.4.1	Monitoring and measurement
4.4.2	Nonconformance and corrective/preventive action
4.4.3	Records
4.4.4	Environmental management system audit
4.5	**Management Review**
Annex A:	Guidance on use of the specification
Annex B:	Bibliography
Annex C:	Links between ISO 14001 and ISO 9000

- Maintained?
- Co-communicated to all employees?

f. Made available to all stakeholders including the general public?

4.2 Planning

4.2.1 Environmental Aspects

5. Has the organization established a procedure to identify the environment aspects of its processes, products, activities, and/or services?
6. Does the organization properly maintain all aspects of its procedure?
7. Does the organization consider all instances of environmental impact when setting environmental objectives?
8. Does the organization keep its information relating to environmental impact up-to-date?

4.2.2 Legal and Other Requirements

9. Has the organization established a procedure to identify legal requirements that apply to its processes, products, activities, and/or services?
10. Has the organization established a procedure for gaining access to all legal requirements identified?

11.   Does the organization properly maintain these procedures?

4.2.3   Objectives and Targets

12.   Has the organization established documented environmental objectives at all relevant functional levels?
13.   Does the organization properly maintain these documented environmental objectives? (For example, are both the objectives and the documents related to them kept up-to-date?)
14.   When establishing/updating its environmental objectives, does the organization consider the following factors:
      a. Legal requirements?
      b. Environmental aspects?
      c. Technological options?
      d. Financial requirements?
      e. Operational requirements?
      f. Business requirements?
      g. Views of all stakeholders?
15.   Are the organization's objectives consistent with the environmental policy?

4.2.4   Environmental Management Program(s)

16.   Has the organization established programs for achieving its environmental objectives?
17.   Do the organization's programs include the following:
      a. Designation of the responsible party for environmental objectives at each functional level?
      b. Means by which the objectives will be accomplished?
      c. Timeframe within which each objective is to be accomplished?
18.   Does the organization amend its programs to include new products, processes, activities, and/or services as they are introduced?

4.3   Implementation and Operation

4.3.1   Structure and Responsibility

19.   Has the organization clearly defined roles, responsibilities, and authority relating to environmental management?
20.   Has the organization documented roles, responsibilities, and authority?
21.   Has the organization communicated roles, responsibilities, and authority to all employees and all other stakeholders?
22.   Does the organization commit the resources necessary to implement, operate, and maintain the EMS? (For example, human, financial, and technological resources)
23.   Has the organization appointed a specific management representative with responsibility and authority for the following:
      a. Ensuring that all EMS requirements are met in accordance with the standard?
      b. Keeping top management informed concerning the performance of the EMS so that performance can be improved continually?

4.3.2   Training, Awareness, and Competence

24.   Has the organization identified the training needs at all of its functional levels?
25.   Does the organization ensure that all employees whose work may affect the environment receive the appropriate training?
26.   Has the organization established procedures to make all employees at all levels in all relevant functions aware of the following:
      a. Importance of positive voluntary compliance with the environmental policy and procedures and all requirements of the EMS?

    b. The environmental aspects of their work activities and the environmental benefits of improving their individual performance?

    c. Their roles and responsibilities in helping the organization comply with its environmental policy and EMS?

    d. The potential negative consequences of failing to follow specified operating procedures?

27. Does the organization properly maintain these procedures?

28. Does the organization ensure that all personnel performing tasks that are potentially hazardous to the environment have the appropriate education, training, and/or experience?

4.3.3 Communication

29. Has the organization established procedures for the following:

    a. Ensuring effective communication among its various levels and functions?

    b. Ensuring effective communication with external stakeholders about environmental concerns?

30. Has the organization recorded its decision concerning how to handle external communication about its significant environmental aspects?

4.3.4 Environmental Management System Documentation

31. Has the organization developed information in an electronic or hardcopy format that does the following?

    a. Describes the core elements of its EMS and how these elements interact?

    b. Gives direction to other documentation related to the EMS?

32. Is this information properly maintained?

4.3.5 Document Control

33. Has the organization established procedures for controlling documents required by ISO 14001?

34. Does the organization ensure that all documents meet the following criteria:

    a. Can documents be located?

    b. Are documents regularly reviewed, updated, and approved by the appropriate authority?

    c. Are up-to-date documents available at all locations where essential functions are performed?

    d. Are outdated documents promptly removed to ensure against their use?

    e. Are outdated documents that must be retained for legal or historical purposes clearly marked for ease of identification?

35. Does the organization ensure that its documents are:

    a. Legible?

    b. Are regularly reviewed, updated, and approved by the appropriate authority?

    c. Readily identifiable?

    d. Maintained in an orderly manner?

    e. Retained for a specified period?

36. Has the organization established procedures for the creation and modification of its documents?

37. Has the organization assigned responsibility for carrying out the procedures for creation and modification of its documents?

38. Does the organization properly maintain all of its procedures relating to its documents?

4.3.6 Operational Control

39. Has the organization identified its operations and activities that are associated with its environmental aspects and tied them to policy and objectives?

40. Has the organization planned in such a way that operational control activities are carried out under specified conditions by doing the following:
    a. Establishing/maintaining operational controls that ensure against deviations from the environmental policy and/or objectives?
    b. Building operational criteria into the procedures?
    c. Establishing/maintaining operational procedures relating to significant environmental impacts and communicating them to suppliers and contractors?

4.3.7   Emergency Preparedness and Response

41. Has the organization established procedures for identifying potential accidents, incidents, and emergency situations?
42. Has the organization established procedures for responding to accidents, incidents, and emergency situations?
43. Has the organization established procedures for preventing/mitigating the environmental effects of accidents, incidents, and emergency situations?
44. Does the organization properly maintain its procedures relating to the environmental aspects of accidents, incidents, and emergency situations?
45. Does the organization periodically review/revise its procedures relating to emergency preparedness and response?
46. Does the organization periodically test (wherever possible) its procedures relating to emergency preparedness and response?

4.4   Checking and Corrective Action

4.4.1   Monitoring and Measurement

47. Has the organization established procedures for monitoring/measuring its operations that may affect the environment?
48. Do the monitoring/measurement procedures include the following:
    a. Recording of information for tracking performance?
    b. Relevant operational controls?
    c. Compliance with environmental objectives?
49. Does the organization properly calibrate its monitoring equipment?
50. Does the organization properly maintain its monitoring equipment?
51. Does the organization make/retain records of calibration and maintenance activities?
52. Has the organization established documented procedures for periodically assessing its compliance with all applicable environmental legislation /regulations?
53. Are the documented procedures properly maintained?

4.4.2   Nonconformance and Corrective/Preventive Action

54. Has the organization established documented procedures for defining responsibility/authority for the following:
    a. Handling/investigating nonconformance?
    b. Taking action to mitigate the negative effects of nonconformance?
    c. Initiating and taking corrective/preventive action?
55. Does the organization ensure that all corrective/preventive action taken is:
    a. Appropriate to the magnitude of the problem?
    b. Commensurate with the environmental problem in question?
56. Does the organization implement changes in documented procedures that result from corrective/preventive action?
57. Does the organization record changes in documented procedures so that the procedures are up-to-date?

4.4.3   Records

58. Has the organization established documented procedures for the following:
    a. Identification of environmental records?

      b. Maintenance of environmental records?

      c. Disposition of environmental records?

59. Are the organization's documented procedures properly maintained?

60. Do the organization's records include the following:

      a. Training records?

      b. Audits?

      c. Reviews?

61. Does the organization ensure that its environmental records:

      a. Are legible?

      b. Are identifiable and traceable to the applicable activity, product, or service?

62. Are the organization's environmental records stored in such a way that they are:

      a. Readily retrievable?

      b. Protected against damage, deterioration, or loss?

63. Has the organization established and recorded retention times for its environmental records?

64. Does the organization maintain its environmental records in a way that:

      a. Is appropriate to the system?

      b. Is appropriate to the organization?

      c. Conforms to the requirements of ISO 14000?

4.4.4 Environmental Management System Audit

65. Has the organization established procedures for periodically auditing its EMS?

66. Do the audit procedures do the following:

      a. Determine whether the EMS conforms to the ISO 14000 Standard and is properly maintained?

      b. Provide audit results to the organization's executive management team?

67. Are the audit procedures/schedule based on the environmental importance of the activities in question and on the results of previous audits?

68. Are the audit procedures sufficiently comprehensive that they cover:

      a. The audit scope?

      b. Frequency and method?

      c. Responsibilities/requirements for conducting audits?

      d. Responsibilities/requirements for reporting results?

4.5 Management Review

69. Does the organization's executive management team periodically review the EMS?

70. Does the organization's management review process ensure that the information needed to properly evaluate the EMS is collected?

71. Are all management reviews properly documented?

APPLICATION SCENARIOS

1. If you were asked to give a seminar on how organizations can reduce hazardous waste, what would you tell the audience? Develop an annotated outline of your seminar.

2. Your employer is a petrochemical processing company. Develop a comprehensive waste reduction plan for the company.

3. How would you go about conducting a waste reduction audit for your college or school? Explain each step and list what your audit would include.

4. Develop a rationale in 500 words or less (typed and double-spaced) to convince your CEO that your company should pursue ISO 14000 registration.

5. Apply the list of criteria (4.0 through 4.5) to the organization of your choice. Use all "No" responses to produce a gap analysis for helping the organization prepare for ISO 14000 registration.

ENDNOTES

1. Main, J. "Here Comes the Big New Clean-Up," *Fortune,* November 21, 1988, Vol. 118, No. 12, p. 102.
2. Ibid.
3. Ibid.
4. LaBar, G. "Seeing Past the Clean Air Act," *Occupational Hazards*, March 1991, p. 29.
5. Cascio, Joseph (ed.). *The ISO 14000 Handbook* (Fairfax, VA: CEEM Information Services, 1996), p. 4.
6. Ibid., pp. 9–10.
7. Tibor, Tom. *ISO 14000: A Guide to the New Environmental Management Standards* (Chicago: IRWIN Professional Publishing, 1996), pp. 171–76.
8. ISO DIS 14001—*Environmental Management Systems—Specification with Guidance for Use* (Chicago: American Society for Quality Control, 1996).
9. Ibid.

CHAPTER SIXTEEN

Violence in the Workplace

Almost one million people are injured or killed in workplace-violence incidents every year in the United States, and the number of incidents is on the rise. In fact, according to the U.S. Department of Justice, the workplace is the most dangerous place to be in the United States.[1] Clearly, workplace violence is an issue of concern to safety and health professionals.

OCCUPATIONAL SAFETY AND WORKPLACE VIOLENCE: THE RELATIONSHIP

The prevention of workplace violence is a natural extension of the responsibilities of safety and health professionals. Hazard analysis, records analysis and tracking, trend monitoring, incident analysis, and prevention strategies based on administrative and engineering controls are all fundamental to both concepts. In addition, emergency response and employee training are key elements of both. Consequently, occupational safety and health professionals are well suited to add the prevention of workplace violence to their normal duties.

SIZE OF THE PROBLEM

Violence in the workplace no longer amounts to just isolated incidents that are simply aberrations. In fact, workplace violence should be considered a common hazard worthy of the attention of safety and health professionals. In a report on the subject, the U.S. Department of Justice revealed the following information:[2]

- About 1,000,000 individuals are the direct victims of some form of violent crime in the workplace every year. This represents approximately 15 percent of all violent crimes committed annually in America. Approximately 60 percent of these violent crimes were categorized as *simple assaults* by the Department of Justice.

- Of all workplace violent crimes reported, over 80 percent were committed by males; 40 percent were committed by complete strangers to the victims; 35 percent by casual acquaintances, 19 percent by individuals well known to the victims, and 1 percent by relatives of the victims.
- More than half of the incidents (56 percent) were not reported to police, although 26 percent were reported to at least one official in the workplace.
- In 62 percent of violent crimes, the perpetrator was not armed; in 30 percent of the incidents, the perpetrator was armed with a handgun.
- In 84 percent of the incidents, there were no reported injuries; 10 percent required medical intervention.
- More than 60 percent of violent incidents occurred in private companies, 30 percent in government agencies, and 8 percent to self-employed individuals.
- It is estimated that violent crime in the workplace caused 500,000 employees to miss 1,751,000 days of work annually, or an average of 3.5 days per incident. This missed work equates to approximately $55,000,000 in lost wages.

LEGAL CONSIDERATIONS

Most issues relating to safety and health have legal ramifications, and workplace violence is no exception. The legal aspects of the issue revolve around the competing rights of violent employees and their co-workers. These conflicting rights create potential liabilities for employers.

Rights of Violent Employees

It may seem odd to be concerned about the rights of employees who commit violent acts on the job. After all, logic suggests that in such situations the only concern would be the protection of other employees. However, even violent employees have rights. Remember, the first thing that law enforcement officers must do after taking criminals into custody is to read them their rights.

This does not mean that an employer cannot take the immediate action necessary to prevent a violent act or the recurrence of such an act. In fact, failure to act prudently in this regard can subject an employer to charges of negligence. However, before taking long-term action that will adversely affect the violent individual's employment, employers should follow applicable laws, contracts, policies, and procedures. Failure to do so can serve to exacerbate an already difficult situation.

Employer Liability for Workplace Violence

Having to contend with the rights of both violent employees and their co-workers, employers often feel as if they are caught between a rock and a hard place. Fortunately, the situation is less bleak than it may first appear due primarily to the exclusivity provision of workers' compensation laws. This provision makes workers' compensation the employee's exclusive remedy for injuries that are work-related. This means that even in cases of workplace violence, as long as the violence occurs within the scope of the victim's employment, the employer is protected from civil lawsuits and the excessive jury verdicts that have become so common.

The key to enjoying the protection of the exclusivity provision of workers' compensation laws lies in determining that violence-related injuries are within the scope of the victim's employment—a more difficult undertaking than one might expect. For example, if the violent act occurred at work but resulted from a non-work-related dispute, does the exclusivity provision apply? What if the dispute was work-related, but the violent act occurred away from the workplace?

Making Work-Related Determinations

The National Institute for Occupational Safety and Health (NIOSH) developed the following guidelines for categorizing an injury as being work-related:

- If the violent act occurred on the employer's premises, it is considered an on-the-job event if one of the following criteria apply:

    The victim was engaged in work activity, apprenticeship, or training.

    The victim was on break, in hallway, restrooms, cafeteria, or storage areas.

    The victim was in the employer's parking lots while working, arriving at, or leaving work.

- If the violent act occurred off the employer's premises, it is still considered an on-the-job event, if one of the following criteria apply:

    The victim was working for pay or compensation at the time, including working at home.

    The victim was working as a volunteer, emergency services worker, law enforcement officer, or firefighter.

    The victim was working in a profit-oriented family business, including farming.

    The victim was traveling on business, including to and from customer–business contacts.

    The victim was engaged in work activity in which the vehicle is part of the work environment (e.g., taxi driver, truck driver, and so on).[3]

RISK REDUCTION STRATEGIES

Figure 16-1 is a checklist that can be used by employers to reduce the risk of workplace violence in their facilities. Most of these risk reduction strategies grow out of the philosophy of *Crime Reduction through Environmental Design,* or CRTED.[4] CRTED has the following four major elements, to which the author has added a fifth (administrative controls).

- Natural surveillance
- Control of access
- Establishment of territoriality
- Activity support
- Administrative controls

Natural Surveillance

This strategy involves designing, arranging, and operating the workplace in a way that minimizes secluded areas. Making all areas inside and outside of the facility easily observable allows for natural surveillance.

Control of Access

One of the most common occurrences of workplace violence involves an outsider entering the workplace and harming employees. The most effective way of stopping this type of incident is to control access to the workplace. Channeling the flow of outsiders to an access-control station, requiring visitor's passes, issuing access badges to

Checklist for Workplace-Violence Risk Reduction

✓ Identify high-risk areas and make them visible. Secluded areas invite violence.

✓ Install good lighting in parking lots and inside all buildings.

✓ Minimize the handling of cash by employees and the amount of cash available on the premises.

✓ Install silent alarms and surveillance cameras where appropriate.

✓ Control access to all buildings (employee badges, visitor check-in/out procedure, visitor passes, and so on).

✓ Discourage working alone, particularly late at night.

✓ Provide training in conflict resolution as part of a mandatory employee orientation.

✓ Conduct background checks before hiring new employees.

✓ Train employees how to handle themselves/respond when a violent act occurs on the job.

✓ Develop policies that establish ground rules for employee behavior/responses in threatening or violent situations.

✓ Nurture a positive, harmonious work environment.

✓ Encourage employees to report suspicious individuals and activities or potentially threatening situations.

✓ Deal with allegations of harassment or threatened violence promptly before the situation escalates.

✓ Take threats seriously and act appropriately.

✓ Adopt a *zero-tolerance* policy toward threatening or violent behavior.

✓ Establish a *violence hot line* so that employees can report potential problems anonymously.

✓ Establish a *threat-management team* with responsibility for preventing and responding to violence.

✓ Establish an *emergency response team* to deal with the immediate trauma of workplace violence.

Figure 16-1
Checklist for workplace-violence risk reduction.

employees, and isolating pickup and delivery points can minimize the risk of violence perpetrated by outsiders.

Establishment of Territoriality

This strategy involves giving employees control over the workplace. With this approach, employees move freely within their established territory but are restricted in other areas. Employees come to know everyone who works in their territory and can, as a result, immediately recognize anyone who shouldn't be there.

Activity Support

Activity support involves organizing workflow and natural traffic patterns in ways that maximize the number of employees conducting natural surveillance. The more employees observing the activity in the workplace, the better.

Administrative Controls

Administrative controls consist of management practices that can reduce the risk of workplace violence. These practices include establishing policies, conducting background checks, and providing training for employees.

OSHA'S VOLUNTARY GUIDELINES

The U.S. Department of Labor, working through the Occupational Safety and Health Administration (OSHA), has established *advisory* guidelines relating to workplace violence.[5] Two key points to understand about these guidelines are as follows:

- The guidelines are *advisory* in nature and *informational* in content. The guidelines *do not* add to or enhance in any way the requirements of the General Duty clause of the OSHAct.
- The guidelines were developed with night retail establishments in mind. Consequently, they have a service-oriented emphasis. However, much of the advice contained in the guidelines can be adapted for use in a manufacturing, processing, and other settings.

Figure 16-2 is a checklist of those elements of the specifications that have broader applications. Any management program relating to safety and health in the workplace should have at least these four elements.

Management Commitment and Employee Involvement

Management commitment and employee involvement are fundamental to developing and implementing any safety program, but they are especially important when trying to prevent workplace violence. The effectiveness of a workplace-violence prevention program may be a life-or-death proposition. Figure 16-3 is a checklist that explains what management commitment means in practical terms. Figure 16-4 describes the practical application of employee involvement. Figure 16-5 is a checklist that can be used to ensure that violence prevention becomes a standard component of organizational plans and operational practices. These three checklists can be used by any type of organization to operationalize the concepts of management commitment and employee involvement.

Workplace Analysis

Workplace analysis is the same process used by safety and health professionals to identify potentially hazardous conditions unrelated to workplace violence. Worksite analysis should be ongoing and have at least the following four components:

- *Records monitoring and tracking.* The purpose of records monitoring and tracking is to identify and chart all incidents of violence and threatening behavior that have

Figure 16-2
Broadly applicable elements of OSHA's advisory guidelines on workplace violence.

> **Checklist of Applicable Advisory Guidelines**
> ✓ Management commitment and employee involvement
> ✓ Worksite analysis
> ✓ Hazard prevention and control
> ✓ Safety and health training

Checklist of Elements of Management Commitment

✓ Hands-on involvement of executive management in developing and implementing prevention strategies.

✓ Sincere, demonstrated concern for the protection of employees.

✓ Balanced commitment to both employees and customers.

✓ Inclusion of safety, health, and workplace-violence prevention in the job descriptions of all executives, managers, and supervisors.

✓ Inclusion of safety, health, and workplace-violence prevention criteria in the performance evaluations of all executives, managers, and supervisors.

✓ Assignment of responsibility for providing coordination and leadership for safety, health, and workplace-violence prevention to a management-level employee.

✓ Provision of the resources needed to prevent workplace violence effectively.

✓ Provision of or guaranteed access to appropriate medical counseling and trauma-related care for employees affected physically and/or emotionally by workplace violence.

✓ Implementation, as appropriate, of the violence-prevention recommendations of committees, task forces, and safety professionals.

Figure 16-3
Elements of management commitment to workplace-violence prevention.

occurred within a given time frame. Records to analyze include the following: incident reports, police reports, employee evaluations, and letters of reprimand. Of course, individual employees' records should be analyzed in confidence by the human resources member of the team. The type of information that is pertinent includes the following:

Where specifically did the incident occur?

What time of day or night did the incident occur?

Prevention Checklist for Employee Involvement

✓ Staying informed concerning all aspects of the organization's safety, health, and workplace-violence program.

✓ Voluntarily complying—in both letter and spirit—with all applicable workplace-violence prevention strategies adopted by the organization.

✓ Making recommendations—through proper channels—concerning ways to prevent workplace violence and other hazardous conditions.

✓ Prompt reporting of all threatening or potentially threatening situations.

✓ Accurate and immediate reporting of all violent and/or threatening incidents.

✓ Voluntary participation on committees, task forces, and/or focus groups concerned with preventing workplace violence.

✓ Voluntary participation in seminars, workshops, and/or other educational programs relating to the prevention of workplace violence.

Figure 16-4
Elements of employee involvement in workplace-violence prevention.

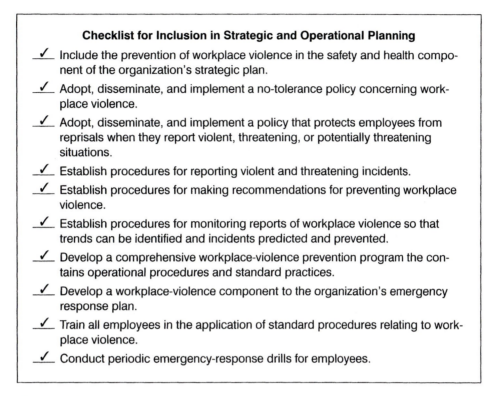

Checklist for Inclusion in Strategic and Operational Planning

✓ Include the prevention of workplace violence in the safety and health component of the organization's strategic plan.

✓ Adopt, disseminate, and implement a no-tolerance policy concerning workplace violence.

✓ Adopt, disseminate, and implement a policy that protects employees from reprisals when they report violent, threatening, or potentially threatening situations.

✓ Establish procedures for reporting violent and threatening incidents.

✓ Establish procedures for making recommendations for preventing workplace violence.

✓ Establish procedures for monitoring reports of workplace violence so that trends can be identified and incidents predicted and prevented.

✓ Develop a comprehensive workplace-violence prevention program the contains operational procedures and standard practices.

✓ Develop a workplace-violence component to the organization's emergency response plan.

✓ Train all employees in the application of standard procedures relating to workplace violence.

✓ Conduct periodic emergency-response drills for employees.

Figure 16-5
Checklist for incorporating workplace-violence prevention in strategic plans and operational practices.

Was the victim an employee? Customer? Outsider?

Was the incident the result of a work-related grievance? Personal?

■ *Trend monitoring/incident analysis.* The use of trend monitoring and incident analysis may prove helpful in determining patterns of violence. If there have been enough incidents to create one or more graphs, the team will want to determine if the graphs suggest a trend or trends. If the organization has experienced only isolated incidents, the team may want to monitor national trends. By analyzing both local and national incidents, the team can generate information that will be helpful in predicting and, thereby, preventing workplace violence. The team should look for trends in severity, frequency, and type of incidents.

■ *Employee surveys/focus groups.* Employees are one of the best sources of information concerning workplace hazards. This is also true when it comes to identifying vulnerabilities to workplace violence. Employee input should be solicited periodically through either written surveys or focus groups or both. Where are we vulnerable? What practices put our employees at risk? These are the types of questions that should be asked of employees. An effective strategy for use with focus groups is to give participants case studies of incidents that occurred in other organizations. Then ask such questions as, "Could this happen here? Why? Or why not? How can we prevent such incidents from occurring here?"

■ *Security analysis.* Is the workplace secure, or could a disgruntled individual simply walk in and harm employees? It is important to ask this question. The team should periodically perform a security analysis of the workplace to identify conditions, situations, procedures, and practices that make employees vulnerable. The types of questions to ask include the following:

Are there physical factors about the facility that make employees vulnerable (e.g., isolated, poorly lighted, infrequently trafficked, or unobservable)?

Is there a process for handling disgruntled customers? Does it put employees at risk?

Are the prevention strategies already implemented working?

Is the training provided to employees having a positive effect? Is more training needed? Who needs the training? What kind of training is needed?

Are there situations in which employees have substantial amounts of money in their possession, on or off-site?

Are there situations in which employees are responsible for highly valuable equipment or materials late at night and/or at isolated locations?

Hazard Prevention and Control

Once hazardous conditions have been identified, the strategies and procedures necessary to eliminate them must be put in place. The two broad categories of prevention strategies are engineering controls and administrative controls, just as they are with other safety and health hazards. In addition to these, organizations should adopt post-incident response strategies as a way to prevent future incidents.

Engineering Controls

Engineering controls relating to the prevention of workplace violence serve the same purpose as engineering controls relating to other hazards. They either remove the hazard, or they create a barrier between it and employees. Engineering controls typically involve changes to the workplace. Examples of engineering controls include the following:

- Installing devices and mechanisms that give employees a complete view of their surroundings (e.g., mirrors, glass or clear plastic partitions, interior windows, and so on)
- Installing surveillance cameras and television screens that allow for monitoring of the workplace
- Installing adequate lighting, particularly in parking lots
- Pruning shrubbery and undergrowth outside and around the facility
- Installing fencing so that routes of egress and ingress to company property can be channeled and, as a result, better controlled
- Arranging outdoor sheds, storage facilities, recycling bins, and other outside facilities for maximum visibility

Administrative Controls

Whereas engineering controls involve making changes to the workplace, administrative controls involve making changes to how work is done. This amounts to changing work procedures and practices. Administrative controls fall into four categories as follows:

- Proper work practices are those that minimize the vulnerability of employees. For example, if a driver has to make deliveries in a high crime area, the company may employ a security guard to go along, change delivery schedules to daylight hours only, or both.
- Monitoring and feedback ensures that proper work practices are being used and that they are having the desired effect. For example, say, a company established a controlled access system in which visitors must check in at a central location and receive a visitor's pass. Is the system being used? Are all employees sticking to specified procedures? Has unauthorized access to the workplace been eliminated?
- Adjustments and modifications are made to violence prevention practices if it becomes clear from monitoring and feedback that they are not working or that improvements are needed.

■ Enforcement involves applying meaningful sanctions when employees fail to follow the established and proper work practices. An employee who has been fully informed concerning a given administrative control, has received the training needed to practice it properly, but consciously decides not to follow the procedure should be disciplined appropriately.

Post-Incident Response

Post-incident response relating to workplace violence is the same as post-incident response relating to traumatic accidents. The first step is to provide immediate medical treatment for injured employees. The second step involves providing psychological treatment for traumatized employees. This step is even more important in cases of workplace violence than with accidents. Employees who are present when a violent incident occurs in the workplace, even if they don't witness it, can experience the symptoms of psychological trauma shown in Figure 16-6. Employees experiencing such symptoms or any others growing out of psychological trauma should be treated by professionals such as psychologists, psychiatrists, clinical nurse specialists, or certified social workers. In addition to one-on-one counseling, employees may also be enrolled in support groups. The final aspect of post-incident response is the investigation, analysis, and report. In this step, safety professionals determine how the violent incident occurred and how future incidents may be prevented, just as post-accident investigations are handled.

Training and Education

Training and education are as fundamental to the prevention of workplace violence as they are to the prevention of workplace accidents and health-threatening incidents. A complete safety and health training program should include a comprehensive component covering all aspects of workplace violence (e.g., workplace analysis, hazard prevention, proper work practices, and emergency response). Such training should be provided on a mandatory basis for supervisors, managers, and employees.

Record Keeping and Evaluation

Maintaining accurate, comprehensive, up-to-date records is just as important when dealing with violent incidents as it is when dealing with accidents and nonviolent incidents. By evaluating records, safety personnel can determine how effective their violence prevention strategies are, where deficiencies exist, and what changes need to be made. The types of records that should be kept are as follows:

■ *OSHA log of injury and illness (OSHA 200).* OSHA regulations require inclusion in the Injury and Illness Log of any injury that requires more than first aid, is a lost-time

Figure 16-6
Symptoms of psychological trauma in cases of workplace violence.

> **Checklist of Symptoms of Psychological Trauma**
> ✓ Fear of returning to work
> ✓ Problems in relationships with fellow employees and/or family members
> ✓ Feelings of incompetence
> ✓ Guilt feelings
> ✓ Feelings of powerlessness
> ✓ Fear of criticism by fellow employees, supervisors, and managers

injury, requires modified duty, or causes loss of consciousness. Of course, this applies only to establishments required to keep OSHA logs. Injuries caused by assaults, which are otherwise recordable, also must be included in the log. A fatality or catastrophe that results in the hospitalization of three or more employees must be reported to OSHA within eight hours. This includes those resulting from workplace violence and applies to all establishments.

- *Medical reports.*   Medical reports of all work injuries should be maintained. These records should describe the type of assault (e.g., unprovoked sudden attack), who was assaulted, and all other circumstances surrounding the incident. The records should include a description of the environment or location, potential or actual cost, lost time, and the nature of injuries sustained.

- *Incidents of abuse.*   Incidents of abuse, verbal attacks, aggressive behavior—which may be threatening to the employee but do not result in injury, such as pushing, shouting, or acts of aggression—should be evaluated routinely by the affected department.

- *Minutes of safety meetings.*   Minutes of safety meetings, records of hazard analyzes, and corrective actions recommended and taken should be documented.

- *Records of all training programs.*   Records of all training programs, attendees, and qualifications of trainers should be maintained.

As part of its overall program, an employer should regularly evaluate its safety and security measures. Top management should review the program regularly, as well as each incident, to determine the program's effectiveness. Responsible parties (managers, supervisors, and employees) should collectively evaluate policies and procedures on a regular basis. Deficiencies should be identified, and corrective action taken. An evaluation program should involve the following activities:

- Establishing a uniform violence reporting system and regular review of reports
- Reviewing reports and minutes from staff meetings on safety and security issues
- Analyzing trends and rates in illness/injury or fatalities caused by violence relative to initial or *baseline* rates
- Measuring improvements based on lowering the frequency and severity of workplace violence
- Keeping up-to-date records of administrative and work practice changes to prevent workplace violence to evaluate their effectiveness
- Surveying employees before and after making job or workplace changes or installing security measures or new systems to determine their effectiveness
- Keeping abreast of new strategies available to deal with violence as they develop
- Surveying employees who experience hostile situations abut the medical treatment they received initially and, again, several weeks afterward, and then several months later
- Complying with OSHA and state requirements for recording and reporting deaths, injuries, and illnesses
- Requesting periodic law enforcement or outside consultant review of the workplace for recommendations on improving employee safety

Management should share violence prevention evaluation reports with all employees. Any changes in the program should be discussed at regular meetings of the safety committee, union representatives, or other employee groups.

CONFLICT RESOLUTION AND WORKPLACE VIOLENCE

When developing a violence prevention program for an organization, the natural tendency is to focus on protecting employees from outsiders. This is important. However, increasingly with workplace violence, the problem is internal. All too often in the modern workplace, conflict between employees is turning violent. Consequently, a violence prevention program is not complete without the following elements: conflict management and anger management.

Conflict Management Component

Disagreements on the job can generate counterproductive conflict. This is one of the reasons why managers in organizations should do what is necessary to manage conflict properly. However, it is important to distinguish between just conflict and counterproductive conflict. Not all conflict is bad. In fact, properly managed conflict that has the improvement of products, processes, people, and/or the work environment as its source is positive conflict.

Counterproductive conflict—the type associated with workplace violence—occurs when employees behave in ways that work against the interests of the overall organization and its employees. This type of conflict is often characterized by deceitfulness, vindictiveness, personal rancor, and anger. Productive conflict occurs when right-minded, well-meaning people disagree, without being disagreeable, concerning the best way to support the organization's mission. Conflict management has the following components:

- Establishing conflict guidelines
- Helping all employees develop conflict prevention/resolution skills
- Helping all employees develop anger management skills

Establishing Conflict Guidelines

Conflict guidelines establish ground rules for discussing and debating differing points of view, differing ideas, and differing opinions concerning how best to accomplish the organization's vision, mission, and broad objectives. Figure 16-7 is an example of an organization's conflict guidelines. Guidelines such as these should be developed with a broad base of employee involvement from all levels in the organization.

Develop Conflict Prevention/Resolution Skills

If managers are going to expect employees to disagree without being disagreeable, they are going to have to ensure that all employees are skilled in the art and science of conflict resolution. The first guideline in Figure 16-7 is an acknowledgment of human nature. It takes advanced human relation skills and constant effort to disagree without being disagreeable. Few people are born with this ability. Fortunately, it can be learned. The following strategies are based on a three-phase model developed by Tom Rusk and described in his book, *The Power of Ethical Persuasion*.[6]

Explore the Other Person's Viewpoint
Allow the other person to present his or her point of view. The following strategies will help make this phase of the discussion more positive and productive.

1. Establish that your goal at this point is mutual understanding.
2. Elicit the other person's complete point of view.
3. Listen nonjudgmentally and do not interrupt.
4. Ask for clarification if necessary.
5. Paraphrase the other person's point of view and restate it to show that you understand.
6. Ask the other person to correct your understanding if it appears to be incomplete.

Explain Your Viewpoint
After you accurately and fully understand the other person's point of view, present your own. The following strategies will help make this phase of the discussion more positive and productive:

Conflict Guidelines
Micro Electronics Manufacturing (MEM)

Micro Electronics Manufacturing encourages discussion and debate among employees at all levels concerning better ways to improve continually the quality of our products, processes, people, and work environment. This type of interaction, if properly handled, will result in better ideas, policies, procedures, practices, and decisions. However, human nature is such that conflict can easily get out of hand, take on personal connotations, and become counterproductive. Consequently, in order to promote productive conflict, MEM has adopted the following guidelines. These guidelines are to be followed by all employees at all levels:

- The criteria to be applied when discussing/debating any point of contention is as follows: Which recommendation is most likely to move our company closer to accomplishing its mission?

- Disagree, but don't be disagreeable. If the debate becomes too hot, stop and give all parties an opportunity to cool down before continuing. Apply your conflict resolution skills and anger management skills. Remember, even when we disagree about how to get there, we are all trying to reach the same destination.

- Justify your point of view by tying it to our mission and require others to do the same.

- In any discussion of differing points of view, ask yourself the following question: "Am I just trying to win the debate for the sake of winning (ego), or is my point of view really the most valid?"

Figure 16-7
Sample conflict guidelines.

1. Ask for the same type of fair hearing for your point of view that you gave the other party.
2. Describe how the person's point of view affects you. Don't point the finger of blame or be defensive. Explain your reactions objectively, keeping the discussion on a professional level.
3. Explain your point of view accurately and completely.
4. Ask the other party to paraphrase and restate what you have said.
5. Correct the other party's understanding if necessary.
6. Review and compare the two positions (yours and that of the other party). Describe the fundamental differences between the two points of view and ask the other party to do the same.

Agree on a Resolution

Once both viewpoints have been explained and are understood, it is time to move to the resolution phase. This is the phase in which both parties attempt to come to an agreement. Agreeing to disagree—in an agreeable manner—is an acceptable solution. The following strategies will help make this phase of the discussion more positive and productive:

1. Reaffirm the mutual understanding of the situation.
2. Confirm that both parties are ready and willing to consider options for coming to an acceptable solution.
3. If it appears that differences cannot be resolved to the satisfaction of both parties, try one or more of the following strategies:

    Take time out to reflect and try again.

    Agree to third-party arbitration or neutral mediation.

Agree to a compromise solution.

Take turns suggesting alternative solutions.

Yield (this time), once your position has been thoroughly stated and is understood. The eventual result may vindicate your position.

Agree to disagree while still respecting each other.

Develop Anger Management Skills

It is difficult, if not impossible, to keep conflict positive when anger enters the picture. If individuals in an organization are going to be encouraged to question, discuss, debate, and even disagree, they must know how to manage their anger. Anger is an intense, emotional reaction to conflict in which self-control may be lost. Anger is a major cause of workplace violence. Anger occurs when people feel that one or more of their fundamental needs are being threatened. These needs include the following:

1. Need for approval
2. Need to be valued
3. Need to be appreciated
4. Need to be in control
5. Need for self-esteem

When one or more of these needs is threatened, a normal human response is to become angry. An angry person can respond in one of four ways:

1. *Attacking.* With this response, the source of the threat is attacked. This response often leads to violence, or at least verbal abuse.

2. *Retaliating.* With this response, you fight fire with fire, so to speak. Whatever is given, you give back. For example, if someone calls your suggestion ridiculous (threatens your need to be valued), you may retaliate by calling his or her suggestion dumb. Retaliation can escalate into violence.

3. *Isolating.* This response is the opposite of venting. With the isolation response, you internalize your anger, find a place where you can be alone, and simmer. The childhood version of this response was to go to your room and pout. For example, when someone fails to even acknowledge your suggestion (threatens your need to be appreciated), you may swallow your anger, return to your office, and boil over in private.

4. *Coping.* This is the only positive response to anger. Coping does not mean that you don't become angry. Rather, it means that, even when you do, you control your emotions instead of letting them control you. A person who copes well with anger is a person who, in spite of his or her anger, stays in control. The following strategies will help employees manage their anger by becoming better at coping:

Avoid the use of anger-inducing words and phrases including the following: *but, you should, you made me, always, never, I can't, you can't,* and so on.

Admit that others don't make you angry; you allow yourself to become angry. You are responsible for your emotions and your responses to them.

Don't let pride get in the way of progress. You don't have to be right every time.

Drop your defenses when dealing with people. Be open and honest.

Relate to other people as equals. Regardless of position or rank, you are no better than they, and they are no better than you.

Avoid the human tendency to rationalize your angry responses. You are responsible and accountable for your behavior.

If employees in an organization can learn to manage conflict properly and to deal with anger positively, the potential for workplace violence will be diminished substantially. Conflict/anger management will not prevent violent acts from outsiders. There are other methods for dealing with outsiders. However, properly managing conflict and anger can protect employees from each other.

EMERGENCY PREPAREDNESS PLAN[7]

To be prepared for properly handling a violent incident in the workplace, employers should form a crisis management team. The team should have only one mission—immediate response to violent acts on the job—and be chaired by a safety and health professional. Team members should receive special training and be updated regularly. The team's responsibilities should be as follows:

- Undergo trauma response training
- Handle media interaction
- Operate telephone/communication teams
- Develop and implement, as necessary, an emergency evacuation plan
- Establish a backup communication system
- Calm personnel after an incident
- Debrief witnesses after an incident
- Ensure that proper security procedures are established, kept up-to-date, and enforced
- Help employees deal with post-traumatic stress
- Keep employees informed about workplace violence as an issue, how to respond when it occurs, and how to help prevent it

=========== APPLICATION SCENARIOS ===========

1. Defend or refute the following statement: If an employer has an employee with a reputation for violence, the employee should be fired immediately. Provide a rationale for your argument.
2. Develop a risk reduction plan that the organization of your choice can use to help decrease the risk of workplace violence.
3. Select an organization with which you are familiar or in which you have an interest. Use the checklist in Figure 16-3 to determine the level of management commitment to the prevention of workplace violence.
4. Develop a report form that any organization can use to make incident reports of violence on the job. Make sure that it provides for all necessary and pertinent information needed for monitoring and tracking violent acts.
5. Select an organization with which you are familiar or in which you have an interest. What engineering controls could this organization implement to help prevent violent accidents?
6. Develop a comprehensive post-incident response checklist any organization can use after a violent act has occurred on the job.

=========== ENDNOTES ===========

1. U.S. Department of Justice. *Violence and Theft in the Workplace* (NCJ-148199) (Annapolis Junction, MD: Bureau of Justice Statistics Clearinghouse, July 1994).
2. U.S. Department of Justice, *Violence and Theft in the Workplace.*
3. National Institute for Occupational Safety and Health and U.S. Department of Health and Human Services. *Homicide in U.S. Workplaces: A Strategy for Prevention and Research* (Washington, D.C.: Centers for Disease Control and Prevention, NIOSH) September 1992.

4. Thomas, Janice L. "A Response to Occupational Violent Crime," *Professional Safety*, June 1997, pp. 27–31.

5. U.S. Department of Labor, Occupational Safety and Health Administration. *Guidelines for Workplace Violence Prevention Programs for Night Retail Establishments*, May 1997.

6. Rusk, Tom. *The Power of Ethical Persuasion* (New York: Penguin Books, 1994), pp. xv–xvii.

7. Illinois State Police. *Do's and Don'ts for the Supervisor*, http://www.state.il.us/isp/viowkplc/vwpp8.htm, August 1996.

Emergency Preparation

Despite the best efforts of all involved, emergencies do sometimes occur. It is important to respond to emergencies in a way that minimizes harm to people and damage to property. To do so requires plans that can be implemented without delay. This chapter provides prospective and practicing safety and health professionals with the information they need to prepare for emergencies in the workplace.

RATIONALE FOR EMERGENCY PREPARATION

An *emergency* is a potentially life-threatening situation, usually occurring suddenly and unexpectedly. Emergencies may be the result of natural and/or human causes. Have you ever witnessed the timely, organized, and precise response of a professional emergency medical crew at an automobile accident? While passers-by and spectators may wring their hands and wonder what to do, the emergency response professionals quickly organize, stabilize, and administer. Their ability to respond in this manner is the result of preparation. Preparation involves a combination of planning, practicing, evaluating, and adjusting to specific circumstances.

When an emergency occurs, immediate reaction is essential. Speed in responding can mean the difference between life and death or between minimal damage and major damage. Ideally, all those involved should be able to respond properly with a minimum of hesitation. This can happen only if all exigencies have been planned for and planned procedures have been practiced, evaluated, and improved.

A quick and proper response—which results because of proper preparation—can prevent panic, decrease the likelihood of injury and damage, and bring the situation under control in a timely manner. Since no workplace is immune to emergencies, preparing for them is critical. An important component of preparation is planning.

EMERGENCY PLANNING AND COMMUNITY RIGHT-TO-KNOW ACT

Title III of the Superfund Amendments and Reauthorization Act of 1986 (SARA) is also known as the Emergency Planning and Community Right-to-Know Act. This law is designed to make information about hazardous chemicals available to a community where they are being used so that residents can protect themselves in the case of an emergency. It applies to all companies that use, make, transport, or store chemicals.

Safety and health professionals involved in developing emergency response plans for their companies should be familiar with the act's requirements for emergency planning. The Emergency Planning and Community Right-to-Know Act includes the four major components discussed in the following paragraphs.

Emergency Planning

The emergency planning component requires that communities form local emergency planning committees (LEPCs) and that states form state emergency response commissions (SERCs). LEPCs are required to develop emergency response plans for the local communities, host public forums, select a planning coordinator for the community, and work with the coordinator in developing local plans. SERCs are required to oversee LEPCs and review their emergency response plans. Plans for individual companies in a given community should be part of that community's larger plan. Local emergency response professionals should use their community's plan as the basis for simulating emergencies and practicing their responses.

Emergency Notification

The emergency notification component requires that chemical spills or releases of toxic substances that exceed established allowable limits be reported to appropriate LEPCs and SERCs. Immediate notification may be verbal as long as a written notification is filed promptly. Such reports must contain at least the following information: (1) the names of the substances released, (2) where the release occurred, (3) when the release occurred, (4) the estimated amount of the release, (5) known hazards to people and property, (6) recommended precautions, and (7) the name of a contact person in the company.

Information Requirements

Information requirements mean that local companies must keep their LEPCs and SERCs and, through them, the public informed about the hazardous substances that the companies store, handle, transport, and/or use. This includes keeping comprehensive records of such substances on file, up to date, and readily available; providing copies of material safety data sheets for all hazardous substances; giving general storage locations for all hazardous substances; providing estimates of the amount of each hazardous substance on hand on a given day; and estimating the average annual amount of hazardous substances kept on hand.

Toxic Chemical Release Reporting

The toxic chemical release reporting component requires that local companies report the total amount of toxic substances released into the environment as either emissions or hazardous waste. Reports go to the Environmental Protection Agency and the state-level environmental agency.

ORGANIZATION AND COORDINATION

A company's emergency response plan should clearly identify the different personnel/groups that respond to various types of emergencies and, in each case, who is in charge. One person should be clearly identified and accepted by all emergency responders as the emergency coordinator. This person should be knowledgeable, at least in a general sense, of the responsibilities of each individual emergency responder and how each relates to those of all other responders. This knowledge must include the order of response for each type of emergency set forth in the plan.

A company's safety and health professional is the obvious person to organize and coordinate emergency responses. However, regardless of who is designated, it is important that (1) one person is in charge, (2) everyone involved knows who is in charge, and (3) everyone who has a role in responding to an emergency is given ample opportunities to practice in simulated conditions that come as close as possible to real conditions.

OSHA STANDARDS

There are no OSHA standards dedicated specifically to the issue of planning for emergencies. However, all OSHA standards are written for the purpose of promoting a safe, healthy, accident-free, and hence emergency-free workplace. Therefore, OSHA standards do play a role in emergency prevention and should be considered when developing emergency plans. For example, exits are important considerations when planning for emergencies. Getting medical personnel in and employees and injured workers out quickly is critical when responding to emergencies. The following sections of OSHA's standards deal exclusively with exit requirements:

Exit arrangements   29 C.F.R. 1910.37(e)

Exit capacity   29 C.F.R. 1910.37(c),(d)

Exit components   29 C.F.R. 1910.37(a)

Exit workings   29 C.F.R. 1910.37(q)

Exit width   29 C.F.R. 1910.37(c)

Exterior exit access 29 C.F.R. 1910.37(g)

A first step for companies developing emergency plans is to review these OSHA standards. This can help safety and health personnel identify and correct conditions that may exacerbate emergency situations before they occur.

FIRST AID IN EMERGENCIES

Workplace emergencies often require a medical response. The immediate response is usually first aid. First aid consists of life-saving measures taken to assist an injured person until medical help arrives.

Since there is no way to predict when first aid may be needed, providing first aid training to employees should be part of preparing for emergencies. In fact, in certain cases, OSHA requires that companies have at least one employee on-site who has been trained in first aid (C.F.R. 1910.151). Figure 17–1 contains a list of the topics that may be covered in a first aid class for industrial workers.

First Aid Training Program

First aid programs are usually available in most communities. The continuing education departments of community colleges and universities typically offer first aid training.

Figure 17–1
Sample course outline for first aid class.

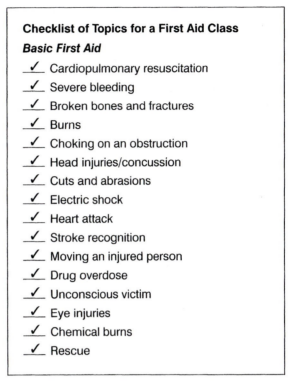

Checklist of Topics for a First Aid Class

Basic First Aid

✓ Cardiopulmonary resuscitation

✓ Severe bleeding

✓ Broken bones and fractures

✓ Burns

✓ Choking on an obstruction

✓ Head injuries/concussion

✓ Cuts and abrasions

✓ Electric shock

✓ Heart attack

✓ Stroke recognition

✓ Moving an injured person

✓ Drug overdose

✓ Unconscious victim

✓ Eye injuries

✓ Chemical burns

✓ Rescue

Classes can often be provided on-site and customized to meet the specific needs of individual companies.

The American Red Cross provides training programs in first aid specifically geared toward the workplace. For more information about these programs, safety and health professionals may contact the national office of the American Red Cross at (202) 639-3200.

The National Safety Council also provides first aid training materials. Its First Aid and Emergency Care Teaching Package contains a slide presentation, overhead transparencies, a test bank, and an instructor's guide. The council also produces a book entitled *First Aid Essentials*. For more information about these materials, safety and health professionals may contact the National Safety Council at (800) 832-0034.

Beyond Training

Training employees in first aid techniques is an important part of preparing for emergencies. However, there is more to being prepared to administer first aid than just training. In addition, it is important to do the following:

1. *Have well-stocked first aid kits available.* First aid kits should be placed throughout the workplace in clearly visible, easily accessible locations. They should be properly and fully stocked and periodically checked to ensure that they stay fully stocked. Figure 17–2 lists the minimum recommended contents for a workplace first aid kit.

2. *Have appropriate personal protective devices available.* With the concerns about AIDS and hepatitis, administering first aid has become more complicated than in the past. The main concerns are with bleeding and other body fluids. Consequently, a properly stocked first aid kit should contain rubber surgical gloves and facemasks or mouthpieces for CPR.

3. *Post emergency telephone numbers.* The advent of 911 service has simplified the process of calling for medical care, police, or fire-fighting assistance. If 911 services

---

Checklist of Contents for a First Aid Kit

✓ Sterile gauze dressings (individually wrapped)

✓ Triangular bandages

✓ Roll of gauze bandages

✓ Assorted adhesive bandages

✓ Adhesive tape

✓ Absorbent cotton

✓ Sterile saline solution

✓ Mild antiseptic for minor wounds

✓ Ipecac syrup to induce vomiting

✓ Powdered activated charcoal to absorb swallowed poisons

✓ Petroleum jelly

✓ Baking soda (bicarbonate of soda)

✓ Aromatic spirits of ammonia

✓ Scissors

✓ Tweezers

✓ Needles

✓ Sharp knife or stiff-backed razor blades

✓ Medicine dropper (eye dropper)

✓ Measuring cup

✓ Oral thermometer

✓ Rectal thermometer

✓ Hot water bag

✓ Wooden safety matches

✓ Flashlight

✓ Rubber surgical gloves

✓ Face masks or mouthpieces

Figure 17–2
Minimum recommended contents of workplace first aid kits.

are not available, emergency numbers for ambulance, hospital, police, fire department, LEPC, and appropriate internal personnel should be posted at clearly visible locations near all telephones in the workplace.

4. *Keep all employees informed.* Some companies require all employees to undergo first aid training; others choose to train one or more employees in each department. Regardless of the approach used, it is important that all employees be informed and keep up to date concerning basic first aid information.

HOW TO PLAN FOR EMERGENCIES

Developing an emergency action plan (EAP) is a major step in preparing for emergencies. A preliminary step is to conduct a thorough analysis to determine the various types of emergencies that may occur. For example, depending on geography and the types of products and processes involved, a company may anticipate such emergencies as the following: fires, chemical spills, explosions, toxic emissions, train derailments, hurricanes, tornadoes, lightning, floods, earthquakes, or volcanic eruptions.

A company's EAP should be a collection of small plans for each anticipated or potential emergency. These plans should have the following components:

1. *Procedures.* Specific, step-by-step emergency response procedures should be developed for each potential emergency.

2. *Coordination.* All cooperating agencies and organizations and emergency responders should be listed along with their telephone number and primary contact person.

3. *Assignments/responsibilities.* Every person who will be involved in responding to a given emergency should know his or her assignment. Each person's responsibilities should be clearly spelled out and understood. One person may be responsible for conducting an evacuation of the affected area, another for the immediate shutdown of all equipment, and another for telephoning for medical, fire, or other types of emergency assistance. When developing this part of the EAP, it is important to assign a back-up per-

---

Checklist for Emergency Planning

Type of Emergency

_____ Fire   _____ Explosion

_____ Chemical spill   _____ Toxic emission

_____ Train derailment   _____ Hurricane

_____ Tornado   _____ Lightning

_____ Flood   _____ Earthquake

_____ Volcanic eruption

Procedures for Emergency Response

1. Controlling and isolating?
2. Communication?
3 Emergency assistance?
4. First aid?
5. Shut-down/evacuation/protection of workers?
6. Protection of equipment/property?
7. Egress, ingress, exits?
8. Emergency equipment (e.g., fire extinguishers)?
9. Alarms?
10. Restoration of normal operations?

Coordination

1. Medical care providers?
2. Fire service providers?
3. LEPC personnel?
4. Environmental protection personnel?
5. Civil defense personnel (in the case of public evacuations)?
6. Police protection providers?
7. Communication personnel?

Figure 17–3
Emergency planning checklist.

son for each area of responsibility. Doing so will ensure that the plan will not break down if a person assigned a certain responsibility is one of the victims.

4. *Accident prevention strategies.* The day-to-day strategies to be used for preventing a particular type of emergency should be summarized in this section of the EAP. In this way, the strategies can be reviewed, thereby promoting prevention.

5. *Schedules.* This section should contain the dates and times of regularly scheduled practice drills. It is best to vary the times and dates so that practice drills don't become predictable and boring. Figure 17–3 is a checklist that can be used for developing an EAP.

EMERGENCY RESPONSE

An emergency response team (ERT) is a special team that responds "to general and localized emergencies to facilitate personnel evacuation and safety, shut down building services and utilities, as needed, work with responding civil authorities, protect and salvage company property, and evaluate areas for safety prior to reentry."[1] The ERT is typically composed of representatives from several different departments such as the following: maintenance, security, safety and health, production/processing, and medical. The actual composition of the team will depend on the size and type of company in question. The ERT should be contained in the assignments/responsibilities section of the EAP.

Not all ERTs are company-based. Communities also have ERTs for responding to emergencies that occur outside of a company environment. Such teams should be included in a company's EAP in the coordinating organizations section. This is especially important for companies that use hazardous materials.

Another approach to ERTs is the emergency response network (ERN). An ERN is a network of ERTs that covers a designated geographical area and is typically responsible for a specific type of emergency.

DEALING WITH THE PSYCHOLOGICAL TRAUMA OF EMERGENCIES

In addition to the physical injuries and property damage that can occur in emergencies, modern safety and health professionals must also be prepared to deal with potential psychological damage. Psychological trauma among employees involved in workplace disasters is as common as it is among combat veterans. According to Johnson, "Traumatic incidents do not affect only immediate survivors and witnesses. Most incidents result in layers of victims that stretch far beyond those who were injured or killed."[2]

Trauma is psychological stress. It occurs as the result of an event, typically a disaster or some kind of emergency, that is so shocking it impairs a person's sense of security or well-being. Johnson calls trauma response "the normal reactions of normal people to an abnormal event."[3] Traumatic events are typically unexpected and shocking, and they involve the reality and/or threat of death.

Dealing with Emergency-Related Trauma

The typical approach to an emergency can be described as follows: control it, take care of the injured, clean up the mess, and get back to work. Often, the psychological aspect is ignored. This leaves witnesses and other coworkers to deal on their own with the trauma they've experienced. "Left to their own inadequate resources, workers can become ill or unable to function. They may develop resentment toward the organization which can lead to conflicts with bosses and co-workers, high employee turnover—even subconscious sabotage."[4]

It is important to respond to trauma quickly, within 24 hours if possible and within 72 hours in all cases. The purpose of the response is to help employees get back to normal by enabling them to handle what they have experienced. This is best accomplished by a team of people who have had special training. Such a team is typically called the trauma response team (TRT).

Trauma Response Team

A company's trauma response team may consist of safety and health personnel who have undergone special training or fully credentialed counseling personnel, depending on the size of the company. In any case, the TRT should be included in the assignments/responsibilities section of the EAP.

The job of the TRT is to intervene as early as possible, help employees acknowledge what they have experienced, and give them opportunities to express how they feel about it to people who are qualified to help. The *qualified to help* aspect is very important. TRT members who are not counselors or mental health professionals should never attempt to provide care that they are not qualified to offer. Part of the trauma training that safety and health professionals receive involves recognizing the symptoms of employees who need professional care and referring them to qualified care providers.

In working with employees who need to deal with what they have experienced, but are not so traumatized as to require referral for outside professional care, a group approach is best. According to Johnson, the group approach offers several advantages, including the following:

■ It facilitates public acknowledgment of what the employees have experienced.

■ It keeps employees informed, thereby cutting down on the number of unfounded rumors and horror stories that will inevitably make the rounds.

■ It encourages employees to express their feelings about the incident. This alone is often enough to get people back to normal and functioning properly.

■ It allows employees to see that they are not alone in experiencing traumatic reactions (e.g., nightmares, flashbacks, shocking memories, and so on) and that these reactions are normal.[5]

APPLICATION SCENARIOS

1. Assume that your company has just experienced an accidental release of a toxic substance. Outline the emergency notification process that you will use. Use your own community as the affected location.

2. Assume that you have been asked to organize a first aid training course. Collect information from the American Red Cross and the National Safety Council on programs available, costs, materials, and so on.

3. Develop a trauma response plan that any organization can use following a workplace disaster.

4. Develop a comprehensive Emergency Action Plan (EAP) for an organization with which you are familiar or in which you have an interest.

ENDNOTES

1. National Safety Council. *Introduction to Occupational Health and Safety* (Chicago: National Safety Council, 1986), p. 341.

2. Johnson, E. "Where Disaster Strikes," *Safety & Health*, February 1992, Vol. 145, No. 2, p. 29.

3. Ibid., p. 28.

4. Ibid., p. 29.

5. Ibid., p. 30.

Accident Investigation and Reporting

When an accident occurs, it is important that it be investigated thoroughly. The results of a comprehensive accident report can help safety and health professionals pinpoint the cause of the accident. This information can then be used to prevent future accidents, which is the primary purpose of accident investigation.

This chapter gives prospective and practicing safety and health professionals the information they need to conduct thorough, effective accident investigations and prepare comprehensive accident reports.

WHEN TO INVESTIGATE

Of course, the first thing to do when an accident takes place is to implement emergency procedures. This involves bringing the situation under control and caring for the injured worker. As soon as all emergency procedures have been accomplished, the accident investigation should begin. Waiting too long to complete an investigation can harm the results. This is an important rule-of-thumb to remember. Another is that *all* accidents, no matter how small, should be investigated. Evidence suggests that the same factors that cause minor accidents cause major accidents.[1] Further, a near miss should be treated like an accident and investigated thoroughly.

There are several reasons why it is important to conduct investigations immediately. First, immediate investigations are more likely to produce accurate information. Conversely, the longer the time span between an accident and an investigation, the greater the likelihood of important facts becoming blurred as memories fade. Second, it is important to collect information before the accident scene is changed and before witnesses begin comparing notes. Human nature encourages people to change their stories to agree with those of other witnesses.[2] Finally, an immediate investigation is evidence of management's commitment to preventing future accidents. An immediate response shows that management cares.[3]

WHAT TO INVESTIGATE

The purpose of an accident investigation is to collect facts. It is not to find fault. It is important that safety and health professionals make this distinction known to all involved. Fault finding can cause reticence among witnesses who have valuable information to share. Causes of the accident should be the primary focus. The investigation should be guided by the following words: *who, what, when, where, why,* and *how.*

This does not mean that mistakes and breaches of precautionary procedures by workers are not noted. Rather, when these things are noted, they are recorded as facts instead of faults. If fault must be assigned, that should come later, after all the facts are in.

In attempting to find the facts and identify causes, certain questions should be asked, regardless of the nature of the accident. The Society of Manufacturing Engineers recommends using the following questions when conducting accident investigations:

1. What type of work was the injured person doing?
2. Exactly what was the injured person doing or trying to do at the time of the accident?
3. Was the injured person proficient in the task being performed at the time of the accident? Had the worker received proper training?
4. Was the injured person authorized to use the equipment or perform the process involved in the accident?
5. Were there other workers present at the time of the accident? If so, who are they and what were they doing?
6. Was the task in question being performed according to properly approved procedures?
7. Was the proper equipment being used including personal protective equipment?
8. Was the injured employee new to the job?
9. Was the process/equipment/system involved new?
10. Was the injured person being supervised at the time of the accident?
11. Are there any established safety rules or procedures that were clearly not being followed?
12. Where did the accident take place?
13. What was the condition of the accident site at the time of the accident?
14. Has a similar accident occurred before? If so, were corrective measures recommended? Were they implemented?
15. Are there obvious solutions that would have prevented the accident?[4]

The answers to these questions should be carefully recorded. You may find it helpful to dictate your findings into a microcassette recorder. This approach allows you to focus more time and energy on investigating and less on taking written notes.

Regardless of how the findings are recorded, it is important to be thorough. What may seem like a minor unrelated fact at the moment could turn out to be a valuable fact later when all of the evidence has been collected and is being analyzed.

WHO SHOULD INVESTIGATE

Who should conduct the accident investigation? Should it be the responsible supervisor? A safety and health professional? A higher-level manager? An outside specialist? There is no simple answer to this question, and there is disagreement among professional people of good will.

In some companies, the supervisor of the injured worker conducts the investigation. In others, a safety and health professional performs the job. Some companies form an investigative team; others bring in outside specialists. There are several reasons for the various approaches used. Factors considered in deciding how to approach accident investigations include the following:

- Size of the company
- Structure of the company's safety and health program
- Type of accident
- Seriousness of the accident
- Number of times that similar accidents have occurred
- Company's management philosophy
- Company's commitment to safety and health

After considering all of the variables just listed, it is difficult to envision a scenario in which the safety and health professional would not be involved in conducting an accident investigation. If the accident in question is very minor, the injured employee's supervisor may conduct the investigation, but the safety and health professional should at least study the accident report and be consulted regarding recommendations for corrective action.

If the accident is so serious that it has widespread negative implications in the community and beyond, responsibility for the investigation may be given to a high-level manager or corporate executive. In such cases, the safety and health professional should assist in conducting the investigation. If a company prefers the team approach, the safety and health professional should be a member of the team and, in most cases, should chair it. Regardless of the approach preferred by a given company, the safety and health professional should play a leadership role in collecting and analyzing the facts and developing recommendations.

CONDUCTING THE INVESTIGATION

The questions in the previous section summarize what to look for when conducting accident investigations. There are five steps to follow in conducting an accident investigation.[5] These steps are explained in the following paragraphs.

Isolate the Accident Scene

You may have seen a crime scene that had been sealed off by the police. The entire area surrounding such a scene is typically blocked off by barriers or heavy yellow tape. This is done to keep curious onlookers from removing, disturbing, or unknowingly destroying vital evidence. This same approach should be used when conducting an accident investigation. As soon as emergency procedures have been completed and the injured worker has been removed, the accident scene should be isolated until all pertinent evidence has been collected or observed and recorded. Further, nothing but the injured worker should be removed from the scene. If necessary, a security guard should be posted to maintain the integrity of the accident scene. The purpose of isolating the scene is to maintain as closely as possible the conditions that existed at the time of the accident.

Record All Evidence

It is important to make a permanent record of all pertinent evidence as quickly as possible. There are three reasons for this: (1) certain types of evidence may be perishable; (2) the longer an accident scene must be isolated, the more likely it is that evidence will be disturbed, knowingly or unknowingly; and (3) if the isolated scene contains a critical piece of equipment or a critical component in a larger process, pressure will quickly mount to get it back in operation. Evidence can be recorded in a variety of ways including written notes, sketches, photography, videotape, dictated observations, and diagrams. In deciding what to record, a good rule-of-thumb is *if in doubt, record it.* It is better to record too much than to skip evidence that may be needed later after the accident scene has been disturbed.

Photograph and/or Videotape the Scene

This step is actually an extension of the previous step. Modern photographic and video-taping technology has simplified the task of observing and recording evidence. Safety and health professionals should be proficient in the operation of a camera, even if it is just an instant camera, and a videotaping camera.

The advent of the digital camera has introduced a new meaning for the concept of "instant photographs." Using a digital camera in conjunction with a computer, photographs of accident scenes can be viewed immediately and transmitted instantly to numerous different locations. Digital camera equipment is especially useful when photographs of accident scenes in remote locations are needed.

Both still and video cameras should be on hand, loaded, and ready to use immediately should an accident occur. As with the previous step, a good rule-of-thumb in photographing and videotaping is *if in doubt, shoot it.* When recording evidence, it is better to have more shots than necessary than it is to risk missing a vital piece of evidence.

A problem with photographs is that, by themselves, they don't always reveal objects in their proper perspective. To overcome this shortcoming, the National Safety Council recommends the following technique:

> When photographing objects involved in the accident, be sure to identify and measure them to show the proper perspective. Place a ruler or coin next to the object when making a close-up photograph. This technique will help to demonstrate the object's size or perspective.[6]

Identify Witnesses

In identifying witnesses, it is important to compile a witness list. Names on the list should be recorded in three categories: (1) primary witnesses, (2) secondary witnesses, and (3) tertiary witnesses (Figure 18–1). When compiling the witness list, ask employees to provide names of all three types of witnesses.

Interview Witnesses

Every witness on the list should be interviewed, preferably in the following order: primary witnesses first, secondary next, and tertiary last. Once all witnesses have been interviewed, it may be necessary to reinterview witnesses for clarification and/or corroboration. Interviewing witnesses is so specialized a process that the next major section is devoted to it.

INTERVIEWING WITNESSES

The techniques used for interviewing accident witnesses are designed to ensure that the information is objective, accurate, as untainted by the personal opinions and feelings of

Figure 18-1
Categories of accident witnesses.

Checklist of Categories of Accident Witnesses

✓ Primary witnesses are eyewitnesses to the accident.

✓ Secondary witnesses are witnesses who did not actually see the accident happen, but were in the vicinity and arrived on the scene immediately or very shortly after the accident.

✓ Tertiary witnesses are witnesses who were not present at the time of the accident nor afterward but may still have relevant evidence to present (e.g., an employee who had complained earlier about a problem with the machine involved in the accident).

witnesses as possible, and able to be corroborated. For this reason, it is important to understand the *when, where,* and *how* of interviewing the accident witnesses.

When to Interview

Immediacy is important. Interviews should begin as soon as the witness list has been compiled and, once begun, should proceed expeditiously. There are two main reasons for this. The first is that a witness's best recollections will be right after the accident. The more time that elapses between the accident and the interview, the more blurred the witness's memory will become. The second reason for immediacy is the possibility of witnesses comparing notes and, as a result, changing their stories. This is just human nature, but it is a tendency that can undermine the value of testimony given and, in turn, the facts collected. Recommendations based on questionable facts are not likely to be valid.

Where to Interview

The best place to interview is at the accident scene. If this is not possible, interviews should take place in a private setting elsewhere. It is important to ensure that all distractions are removed, interruptions are guarded against, and the witness is not accompanied by other witnesses. All persons interviewed should be allowed to have their say without fear of contradiction or influence by other witnesses or employees. It is also important to select a neutral location in which witnesses will feel comfortable. Avoid the principal's office syndrome by selecting a location that is not likely to be intimidating to witnesses.

How to Interview

The key to getting at the facts is to put the witness at ease and to listen. Listen to what is said, how it is said, and what is not said. Ask questions that will get at the information listed earlier in this chapter, but phrase them in an open-ended format. For example, instead of asking, "Did you see the victim pull the red lever?" phrase your question as follows: "Tell me what you saw." Don't lead witnesses with your questions or influence them with gestures, facial expressions, tone of voice, or any other form of nonverbal communication. Interrupt only if absolutely necessary to seek clarification on a critical point. Remain nonjudgmental and objective.

An accident investigation is similar to a police investigation of a crime in that the information being sought can be summarized as who, what, when, where, why, and how. As information is given, it may be necessary to take notes. If you can keep your note-taking to a minimum during the interview, your chances of getting uninhibited information are increased. Note-taking can distract and even frighten a witness.

An effective technique is to listen during the interview and make mental notes of critical information. At the end of the interview, summarize what you have heard and have the witness verify your summary. After the witness leaves, develop your notes immediately.

A question that sometimes arises is, "Why not tape the interview?" Safety and health professionals disagree on the effectiveness and advisability of taping. Those who favor taping claim it allows the interviewer to concentrate on listening without having to worry about forgetting a key point or having to interrupt the witnesses to jot down critical information. It also preserves everything that is said for the record as well as the tone of voice in which it is said. A complete transcript of the interview also ensures that information is not taken out of context.

Those opposed to taping say that taping devices tend to inhibit witnesses so that they are not as forthcoming as they would be without taping. Taping also slows down the investigation while the taped interview is transcribed and the interviewer wades through voluminous testimony trying to separate critical information from irrelevant information.

In any case, if the interview is to be taped, the following rules-of-thumb should be applied:

- Use the smallest, most unobtrusive taping device available, such as a microcassette recorder.
- Inform the witness that the interview will be taped.
- Make sure the taping device is working properly and that the tape it contains can run long enough so that you don't have to interrupt the witness to change it.
- Take time at the beginning of the interview to discuss unrelated matters long enough to put the witness at ease and overcome the presence of the taping device.
- Make sure the personnel are available to transcribe the tapes immediately.
- Read the transcripts as soon as they are available and highlight critical information.

An effective technique to use with eyewitnesses is to ask them to reenact the accident for you. Of course, the effectiveness of this technique is enhanced if the reenactment can take place at the accident site. However, even when this is not possible, an eyewitness reenactment can yield valuable information.

In using the reenactment technique, a word of caution is in order. If an eyewitness does exactly what the victim did, there may be another accident. Have the eyewitnesses explain what they are going to do before letting them do it. Then, have them *simulate* rather than actually perform the steps that led up to the accident.

REPORTING ACCIDENTS

An accident investigation should culminate in a comprehensive accident report. The purpose of the report is to record the findings of the accident investigation, the cause or causes of the accident, and recommendations for corrective action.

OSHA has established requirements for reporting and record keeping. According to OSHA document 2056,

> Employers of 11 or more employees must maintain records of occupational injuries and illnesses as they occur. Employers with 10 or fewer employees are exempt from keeping such records unless they are selected by the Bureau of Labor Statistics (BLS) to participate in the Annual Survey of Occupational Injuries and Illnesses.[7]

All injuries and illnesses are supposed to be recorded, regardless of severity, if they result in any of the outcomes shown in Figure 18–2. If an accident results in the death of an employee or hospitalization of five or more employees, a report must be submitted to the nearest OSHA office within 48 hours. This rule applies regardless of the size of the company.

Accident report forms vary from company to company. However, the information contained in them is fairly standard. Regardless of the type of form used, an accident report should contain at least the information needed to meet the record-keeping requirements set forth by OSHA. This information includes at least the following, according to the National Safety Council:

Figure 18–2
OSHA record-keeping requirements.

Checklist of OSHA Record-Keeping Requirements

Injuries/illnesses must be recorded if they result in any of the following:

- ✓ Death
- ✓ One or more lost workdays
- ✓ Restriction of motion or work
- ✓ Loss of consciousness
- ✓ Transfer to another job
- ✓ Medical treatment (more than first aid)

ACCIDENT REPORT FORM*

Fairmont Manufacturing Company
1501 Industrial Park Road
Fort Walton Beach, Florida 32548
904-725-4041

Victim-Related Information

Person completing report _____ Case no. _____

Gender _____ Age _____

Date of accident/illness _____

Victim's home address/telephone _____

_____

Victim's assignment at the time of the accident and length of time in that assignment:

_____

_____

_____

Victim's normal job and length of time in that job: _____

_____

_____

Time of injury/illness and phase of victim's workday: _____

_____

_____

Severity of the injury (e.g., hospitalization required, first aid required, etc.): _____

_____

_____

Type of injury and body part(s) injured: _____

_____

Exact location of the accident (which facility, department, place within the department):

_____

_____

Physician and hospital: _____

_____

Figure 18–3
Sample accident report form.
*One form for each injured worker.

- Case number of the accident
- Victim's department or unit
- Location and date of the accident or date that an illness was first diagnosed
- Victim's name, social security number, sex, age, home address, and telephone number
- Victim's normal job assignment and length of employment with the company
- Victim's employment status at the time of the accident (i.e., temporary, permanent, full-time, part-time)
- Case numbers and names of others injured in the accident
- Type of injury and body part(s) injured (e.g., burn to right hand; broken bone, lower right leg) and severity of injury (i.e., fatal, first aid only required, hospitalization required)

Accident-Related Information

Accident description with step-by-step sequence of events: _____

_____

_____

Task and specific activity at the time of the accident: _____

_____

_____

Posture/proximity of employee at the time of the accident: _____

_____

_____

Supervision status at the time of the accident: _____

_____

_____

Apparent causes including conditions, actions, events, and activities and other

contributing factors: _____

_____

_____

Recommendations for corrective action: _____

_____

_____

Case numbers and names of other persons injured in the accident: _____

_____

_____

Witnesses to the accident, and dates/places of their interviews:

_____   _____

_____   _____

_____   _____

_____   _____

_____   _____

Name   Date

Figure 18–3 (continued)

- Name, address, and telephone number of the physician called
- Name, address, and telephone number of the hospital to which the victim was taken
- Phase of the victim's workday when the accident occurred (e.g., beginning of shift, during break, end of shift, and so on)
- Description of the accident and how it took place, including a step-by-step sequence of events leading up to the accident
- Specific tasks and activities with which the victim was involved at the time of the accident (e.g., task: mixing cleaning solvent; activity: adding detergent to the mixture)
- Employee's posture/proximity related to his or her surroundings at the time of the accident (e.g., standing on a ladder; bent over at the waist inside the robot's work envelope)
- Supervision status at the time of the accident (i.e., unsupervised, directly supervised, indirectly supervised)

- Causes of the accident
- Corrective actions that have been taken so far
- Recommendations for additional corrective action[8]

In addition to these items, you may want to record such additional information as the list of witnesses; dates, times, and places of interviews; historical data relating to similar accidents; information about related corrective actions that were made previously but had not yet been followed up on; and any other information that might be relevant. Figure 18–3 is an example of an accident report form that meets the OSHA record-keeping specifications.

APPLICATION SCENARIOS

1. Locate an accident report form from any organization in your community. Based on what you learned in this chapter concerning what to investigate, is the form comprehensive enough? What additional information might be asked?
2. Describe in detail how you would go about investigating one of the following accident scenarios:
   a. Reading a production report while walking, John Jones entered the work envelope of a robot, which broke two of his ribs.
   b. A maintenance worker on a lift two stories above the shop floor dropped a wrench that hit a machinist in the head. The machinist was wearing a hard hat.
   c. A forklift driver dismounted the vehicle to move a box out of his path. He slipped on an oily place and injured his back.
3. For one of the accident examples in Scenario 2, describe in detail how you would interview witnesses.
4. Using the accident you chose in Scenario 3, complete a comprehensive accident report.

ENDNOTES

1. Society of Manufacturing Engineers. *The Manufacturing Engineer's Handbook*, Vol. 5, (Dearborn, MI: Society of Manufacturing Engineers, 1988), pp. 12–21.
2. Ibid.
3. Ibid.
4. Society of Manufacturing Engineers. *The Manufacturing Engineer's Handbook*, pp. 12–21.
5. National Safety Council. *Supervisor's Safety Manual*, p. 71.
6. Ibid.
7. OSHA 2056, 1991 (Revised), U.S. Department of Labor, p. 11.
8. National Safety Council. *Supervisor's Safety Manual*, pp. 76–77.

Promoting Safety

- Company Safety Policy
- Safety Rules and Regulations
- Employee Participation in Promoting Safety
- Suggestion Programs
- Visual Awareness
- Safety Committees
- Incentives
- Competition
- Teamwork Approach to Promoting Safety

The purpose of safety promotion is to keep employees focused on doing their work the safe way, every day. This chapter provides prospective and practicing safety and health professionals with information that will enable them to promote safety effectively.

COMPANY SAFETY POLICY

Promoting safety begins with having a published company safety policy. The policy should make it clear that safe work practices are expected of all employees at all levels at all times. The safety policy serves as the foundation upon which all other promotional efforts are built.

Figure 19–1 is an example of a company safety policy. This policy briefly and succinctly expresses the company's commitment to safety. It also indicates clearly that

Figure 19-1
Sample company safety policy.

Okaloosa Poultry Processing, Inc.

414 Baker Highway
Crestview, Florida 36710

Safety Policy

It is the policy of this company to ensure a safe and healthy workplace for employees, a safe and healthy product for customers, and a safe and healthy environment for the community. OPP, Inc. is committed to safety on the job and off. Employees are expected to perform their duties with this commitment in mind.

---

employees are expected to perform their duties with safety foremost in their minds. With such a policy in place and clearly communicated to all employees, other efforts to promote safety will have solid backing.

A company's safety policy need not be long. In fact, a short and simple policy is better. Regardless of its length or format, a safety policy should convey at least the following messages:

1. The company is committed to safety and health.
2. Employees are expected to perform their duties in a safe and healthy manner.
3. The company's commitment extends beyond the walls of its plant to include customers and the community.

SAFETY RULES AND REGULATIONS

A company's safety policy is translated into everyday action and behavior by rules and regulations. Rules and regulations define behavior that is acceptable and unacceptable from a safety and health perspective. From a legal point of view, an employer's obligations regarding safety rules can be summarized as follows:

1. Employers must have rules that ensure a safe and healthy workplace.
2. Employers must ensure that all employees are knowledgeable about the rules.
3. Employers must ensure that safety rules are enforced objectively and consistently.

The law tends to view employers who do not meet these three criteria as being *negligent*. Having the rules is not enough. Having rules and making employees aware of them is not enough. Employers must develop appropriate rules, familiarize all employees with them, and enforce the rules. It is this final step—enforcement—from which most negligence charges arise.

Figure 19–2 contains guidelines to follow when developing safety rules. These guidelines will help ensure a safe and healthy workplace without unduly inhibiting workers in the performance of their jobs. This is an important point for prospective and practicing safety and health professionals to understand. Fear of negligence charges can influence an employer in such a way that the book of safety rules becomes a multivolume nightmare that is beyond the comprehension of most employees.

Such attempts to avoid costly litigation, penalties, or fines by regulating every move that employees make and every breath that they take are likely to backfire. Remember, employers must do more than just write rules. They must also familiarize all employees with them and enforce them. This will not be possible if the rulebook is as thick as an unabridged dictionary. Apply common sense when writing safety rules.

Figure 19-2
Guidelines for developing safety rules and regulations.

Checklist of Guidelines for Developing Safety Rules

✓ Minimize the number of rules to the extent possible. Too many rules can result in rule *overload*.

✓ Write rules in clear and simple language. Be brief and to the point, avoiding ambiguous or overly technical language.

✓ Write only the rules that are necessary to ensure a safe and healthy workplace. Do not *nitpick*.

✓ Involve employees in the development of rules that apply to their specific areas of operation.

✓ Develop only rules that can and will be enforced.

✓ Use common sense in developing rules.

Jones Petroleum Products, Inc.
Highway 90 East
DeFuniak Springs, Florida 32614

Suggestion Form

Name of employee: _____ Date of suggestion: _____

Department: _____

Suggested improvement: _____

_____

_____

_____

Date logged in: _____ Time: _____

Logged in by: _____

Action taken: _____

_____

Current status: _____

_____

Date of response to employee: _____

Person responding: _____
                                     (Signature)

Figure 19–3
Sample safety suggestion form.

EMPLOYEE PARTICIPATION IN PROMOTING SAFETY

One of the keys to successfully promoting safety is to involve employees. They usually know better than anyone where hazards exist. In addition, they are the ones who must follow safety rules. A fundamental rule of management is *if you want employees to make a commitment, involve them from the start.* One of the most effective strategies for getting employees to commit to the safety program is to involve them in the development of it. This way, your program becomes their program.

SUGGESTION PROGRAMS

Suggestion programs, if properly handled, promote safety and health. Well-run suggestion programs offer two advantages: (1) they solicit input from the people most likely to know where hazards exist; and (2) they involve and empower employees which, in turn, gives them ownership of the safety program.

Suggestion programs must meet certain criteria to be effective:

- All suggestions must receive a formal response.
- All suggestions must be answered immediately.
- Management must monitor the performance of each department in generating and responding to suggestions.
- System costs and savings must be reported.
- Recognition and awards must be handled promptly.
- Good ideas must be implemented.
- Personality conflicts must be minimized.[1]

Figure 19–4
Sample safety reminder sign.

Suggestion programs that meet these criteria are more likely to be successful than those that don't. Figure 19–3 is an example of a suggestion form that may be used as part of a company's safety program. Note that all of the following must be recorded: the date that the suggestion was submitted, the date that the suggestion was logged in, and the date that the employee received a response.

VISUAL AWARENESS

We tend to be a visual society. This is why television and billboards are so effective in marketing promotions. Making a safety and health message visual can be an effective way to get the message across. Figure 19–4 is a sign that gives machine operators a visual reminder to use the appropriate machine guards. Such a sign is placed on or near the machine in question. If operators cannot activate their machines without first reading this sign, they will be reminded to use the safe way every time they operate the machine.

Figure 19–5 may be placed on the door leading into the hard hat area or on a stand placed prominently at the main point of entry if there is no door. Such a sign will help prevent inadvertent slip-ups when employees are in a hurry or are thinking about something else.

Several rules-of-thumb can help ensure the effectiveness of efforts to make safety visual:

- Change signs, posters, and other visual aids periodically. Visual aids left up too long begin to blend into the background and are no longer noticed.
- Involve employees in developing the messages that will be displayed on signs and posters. Employees are more likely to notice and heed their own messages than those of others.
- Keep visual aids simple and the message brief.
- Make visual aids large enough to be seen easily from a reasonable distance.
- Locate visual aids for maximum effect.
- Use color whenever possible to attract attention to the visual aid.

SAFETY COMMITTEES

Another way to promote safety through employee involvement is the safety committee. Safety committees provide a formal structure through which employees and manage-

Figure 19–5
Sample safety reminder sign.

ment can funnel concerns and suggestions about safety and health issues. The composition of the safety committee can be a major factor in the committee's success or failure.

The most effective committees are those that are composed of a broad cross-section of workers representing all departments. This offers two advantages: (1) it gives each member of the committee a constituent group for which he or she is responsible; and (2) it gives all employees a representative voice on the committee.

There is disagreement over whether an executive-level manager should serve on the safety committee. On the one hand, an executive-level participant can give the committee credibility, visibility, and access. On the other hand, the presence of an executive manager can inhibit the free flow of ideas and concerns. The key to whether an executive manager's participation will be positive or negative lies in the personality and management skills of the executive in question.

An executive who knows how to put employees at ease, interact in a nonthreatening manner, and draw people out will add to the effectiveness of the committee. An executive with a threatening attitude will render the committee useless. Consequently, the author recommends the involvement of a very carefully selected executive manager on the safety committee.

The safety and health professional should be a member of the committee serving as an advisor, facilitator, and catalyst. Committee members should select a chairperson from the membership and a recording secretary for taking minutes and maintaining committee records. Neither the executive manager nor the safety and health professional should serve as chairperson, but either can serve as recording secretary. Excluding executive managers and safety and health professionals from the chair will give employees more ownership in the committee.

Safety committees will work only if members are empowered and trained to identify hazards and take steps to eliminate them. Such training can help ensure that safety committee members identify actual problems rather than just deal with symptoms.

INCENTIVES

If properly used, incentives can help promote safety. However, the proper use of incentives is a widely misunderstood concept. To promote safety effectively, incentives must be properly structured. Puffer recommends the following strategies for enhancing the effectiveness of incentive programs:

1. *Define objectives.* Begin by deciding what is supposed to be accomplished by the incentive program.

2. *Develop specific criteria.* On what basis will the incentives be awarded? This question should be answered during the development of the program. Specific criteria define the type of behavior and level of performance that is to be rewarded as well as guidelines for measuring success.

3. *Make rewards meaningful.* For an incentive program to be effective, the rewards must be meaningful to the recipients. Giving an employee a reward that he or she does not value will not produce the desired results. To determine what types of rewards will be meaningful, it is necessary to involve employees.

4. *Recognize that only employees who will participate in an incentive program know what incentives will motivate them.* In addition, employees must feel it is *their* program. This means that employees should be involved in the planning, implementation, and evaluation of the incentive program.

5. *Keep communications clear.* It is important for employees to understand fully the incentive program and all of its aspects. Communicate with employees about the program, ask for continual feedback, listen to the feedback, and act on it.

6. *Reward teams.* Rewarding teams can be more effective than rewarding individuals. This is because work in the modern industrial setting is more likely to be accomplished by a team than an individual. When this is the case, other team members may resent the recognition given to an individual member. Such a situation can cause the incentive program to backfire.[2]

COMPETITION

Competition is another strategy that can be used to promote safety. However, if this approach is not used wisely, it can backfire and do more harm than good. To a degree, most people are competitive. A child's competitive instinct is nurtured through play and reinforced by sports and school activities. Safety and health professionals can use the adult's competitive instinct when trying to motivate employees, but competition on the job should be carefully organized, closely monitored, and strictly controlled. Competition that is allowed to get out of hand can lead to cheating and hard feelings among fellow workers.

Competition can be organized between teams, shifts, divisions, or even plants. Here are some tips that will help safety and health professionals use competition in a positive way while ensuring that it does not get out of hand:

- Involve the employees who will compete in planning programs of competition.
- Where possible, encourage competition among groups rather than individuals, while simultaneously promoting individual initiative within groups.
- Make sure that the competition is fair by ensuring that the resources available to competing teams are equitably distributed and that human talent is as appropriately spread among the teams as possible.

The main problem with using competition to promote safety is that it can induce competing teams to cover up or fail to report accidents just to win. Safety and health professionals should be particularly attentive to this situation and watch carefully for evidence that accidents are going unreported. If this occurs, the best approach is to confront the situation openly and frankly. Employees should be reminded that improved safety is the first priority and winning the competition is second. Failing to report an accident should be grounds for eliminating a team from competition.

TEAMWORK APPROACH TO PROMOTING SAFETY

Increasingly, teamwork is stressed as the best way to get work done in the contemporary workplace. Consequently, it follows that the teamwork approach is an excellent way to promote safety. This section is limited to covering teamwork as it relates specifically to the promotion of safety.

Characteristics of Effective Teams

Effective teams share several common characteristics: supportive environment, team player skills, role clarity, clear direction, team-oriented rewards, and accountability.

Supportive Environment
The characteristics of a team-supportive environment are well known. These characteristics are

- Open communication
- Constructive, nonhostile interaction
- Mutually supportive approach to work
- Positive, respectful climate

Team Player Skills

Team player skills are personal characteristics of individuals that make them good team players. They include the following:

- Honesty
- Selflessness
- Initiative
- Patience
- Resourcefulness
- Punctuality
- Tolerance
- Perseverance

Role Clarity

On any team, different members play different roles. Consider the example of a football team. When the offensive team is on the field, each of the 11 team members has a specific role to play. The quarterback plays one role; the running backs, another; the receivers, another; the center, another; and the lineman, another. Each of these roles is different but important to the team. When each of these players executes his role effectively, the team performs well.

But what would happen if the center suddenly decided he wanted to pass the ball? What would happen if one of the linemen suddenly decided that he wanted to run the ball? Of course, chaos would ensue. A team cannot function if team members try to play roles that are assigned to other team members. Role clarity means that all members understand their respective roles on the team and play those roles.

Clear Direction

What is the team's purpose? What is the team supposed to do? What are the team's responsibilities? These are the types of questions that people ask when they are assigned to teams. The team's charter should answer such questions. The various components of a team's charter are as follows:

1. *Mission.* The team's mission statement defines its purpose and how the team fits into the larger organization. In the case of a safety promotion team, it explains the team's role in the organization's overall safety program.

2. *Objectives.* The team's objectives spell out exactly what the team is supposed to accomplish in terms of the safety program.

3. *Accountability measures.* The team's accountability measures spell out how the team's performance will be evaluated.

Figure 19-6 is an example of a team charter for the safety promotion team in a manufacturing company. This charter clearly defines the committee's purpose, where it fits into the overall organization, what it is supposed to accomplish, and how the committee's success will be measured.

Team-Oriented Rewards

One of the most commonly made mistakes in organizations is attempting to establish a teamwork culture while maintaining an individual-based reward system. If teams are to function fully, the organization must adopt team-oriented rewards, incentives, and recognition strategies. For example, teams function best when the financial rewards of its members are tied at least partially to team performance. Performance appraisals that contain criteria relating to team performance, in addition to individual performance, promote teamwork. The same concept applies to recognition activities.

Figure 19-6
Sample team charter.

Team Charter

Safety Promotion Team
MTC Corporation

Mission
The mission of the *Safety Promotion Team* at MTC Corporation is to make all employees at all levels of the company aware of the importance of safety and health, and, having made them aware, to keep them aware.

Objectives
1. Identify innovative, interesting ways to communicate the company's safety rules and regulations to employees.
2. Develop a company-wide suggestion system to solicit safety-related input from employees.
3. Identify eye-catching approaches for making safety a *visible* issue.
4. Develop appropriate safety competition activities.

Accountability Measures
The quality of participation in all of the activities of this team will be assessed by the team leader and included in the annual performance appraisal of each team member. Team members are expected to be consistent in their attendance, punctual, cooperative, and mutually supportive.

Accountability

There is a rule-of-thumb in management that says, "If you want to improve performance, measure it." Accountability is about being held responsible for accomplishing specific objectives or undertaking specific actions. The most effective teams know what their responsibilities are and how their success will be measured.

Potential Benefits of Teamwork in Promoting Safety

Teamwork can have both direct and indirect benefits for an organization. Through teamwork, counterproductive internal competition and internal politics are replaced by collaboration. When this happens, the following types of benefits typically accrue:

- Better understanding of safety rules/regulations
- Visibility for safety
- Greater employee awareness
- Positive, productive competition
- Continual improvement
- Broader employee input and acceptance

Potential Problems with Teams

Teamwork can yield important benefits, but as with any concept, there are potential problems. The most pronounced potential problems with teams are as follows:

- It can take a concerted effort over an extended period of time to mold a group into an effective team, but a team can fall apart quickly.

- Personnel changes are common in organizations, but personnel changes can disrupt a team and break down team cohesiveness.
- Participative decision making is inherent in teamwork. However, this approach to decision making takes time, and time is often in short supply.
- Poorly motivated and lazy employees can use a team to blend into the crowd, to avoid participation. If one team member sees another slacking, he or she may respond in kind.

These potential problems can be prevented, of course. The first step in doing so is recognizing them. The next step is ensuring that all team members fulfill their responsibilities to the team and to one another.

Responsibilities of Team Members

Accountability in teamwork amounts to team members fulfilling their individual responsibilities to the team and to each other. These responsibilities are as follows:

- Active participation in all team activities
- Punctuality in attendance of meetings
- Honesty and openness toward fellow team members
- Making a concerted effort to work well with team members
- Being a good listener for other team members
- Being open to the ideas of others

If individual team members fulfill these responsibilities to each other and the team, the potential problems with teams can be overcome, and the benefits of teamwork can be fully realized. It is important for members of the safety team to understand these responsibilities, accept them, and set an example of fulfilling them. If this happens, the benefits to the organization will go well beyond just safety and health.

APPLICATION SCENARIOS

1. Develop a comprehensive safety promotion plan that can be replicated by any organization. Your plan should be for an actual organization.
2. Write a letter to a fellow safety student or practitioner explaining how you would use incentives to promote safety in an organization.
3. Identify a company in your community that employs a safety professional and schedule an appointment with that individual. Ask to see the following:
   a. Safety policy
   b. Safety rules and regulations
   c. Visual safety instructions

ENDNOTES

1. McDermott, B. "Employees Are Best Source of Ideas for Constant Improvement," *Total Quality Newsletter,* July–August 1990, Vol. 1, No. 4, p. 5.
2. Puffer, T. "Eight Ways to Construct Effective Service Reward Systems," *Reward & Recognition Supplement, Training,* August 1990, pp. 8–12.

Index